TINNEMAHA

Tinnemaha Ranch

TINNEMAHA

BY

J.R. ALCORN

Fairview West Publishing
Fallon, Nevada 1991

FAIRVIEW WEST PUBLISHING
510 South Maine St., Fallon Nevada 89406

Printed by Braun-Brumfield, Inc.
Ann Arbor, Michigan, U.S.A.

ISBN: 0-9620221-0-7
Library of Congress No.: 91-73674

Books By J.R. (Ray) Alcorn

The Birds of Nevada

Tinnemaha

Reno
Bishop
Big Pine
Fullerton
Durango
Baca County
Greeley
Ionia
Vilas
Stoningto

Monticello

TABLE OF CONTENTS

ACKNOWLEDGEMENTS

I would like to give thanks to my brother, Lloyd Alcorn, and sisters, Velma Watkins and Marie Ruppre, for the many true stories and information of our past.

Special thanks to my cousins Patricia Meyer Franks of Wawaka, Indiana and the late Carroll G. Bryan for many family records. Windy Cox of Big Pine, California who reviewed and made editorial suggestions.

I am grateful to the late Lawson Brainerd of Mill Valley, California for material on his life in Owens Valley, to the late Gladys Shumaker Levering for the information about Fish Springs School and to Dorothy Heterbrink Theaker for her memories of a summer vacation on the Tinnemaha Ranch. My son Albert Alcorn, Katy McGowan, Marla Carey, and Sonya Johnson for reproducing and taking pictures to make this book more complete. Also to Velma Watkins who contributed family photos.

Without those who edited, typed, and assisted in various ways this book would not have been completed. I appreciate the support and interest of all of you: Katy McGowan, Marla Carey, Charlene Miller, Colleen Worlton, Ruth Parks, Nikki Lott and Deborah Alcorn. Thanks to the Inyo Register's Publisher, C. Deane Funk, and General Manager, Pete Doughtie, for permission to research the newspapers for the years 1910 to 1912. The names of my family and friends who contributed to this book will be found in the respective chapters. Thanks to all the people I have not mentioned.

Special thanks to my wife, Capitola "Cappy" Marney Alcorn for helping make the final edit to complete this book.

INTRODUCTION

TINNEMAHA is about the life of William Harrison Alcorn and his family. As a child, he traveled by covered wagon from Kentucky in 1885 to Kansas where there were no roads, only wagon trails. William Harrison later left Kansas to Colorado then to Owens Valley by train.

He purchased property in the Sunland District south of Bishop, California in 1911. Later moving his family to the Red Mountain Fruit Ranch south of Big Pine. The ranch was later named "Tinnemaha."

This book tells of true stories and experiences while living in Owens Valley with Lawson Brainerd, Bill Banta, Gladys Shumaker Levering and Dorothy Heterbrink Theaker, as well as many others.

In 1929, when the Tinnemaha Ranch was sold to the city of Los Angeles, W. H. Alcorn moved his family to Fallon, Nevada. He bought an eighty acre farm four miles west of town and built a brick house similar in size to the Tinnemaha ranch house. The east-west road north of the house was later named Alcorn Road in honor of W. H. Alcorn.

William Harrison Alcorn lived from February 20, 1874, to October 6, 1946. His wife Delpha Konkel Alcorn was born August 8, 1881, and passed away April 11, 1957. The youngest of us boys, Earl Jason Alcorn, was born September 23, 1915, and lived to November 28, 1944. All three are buried in the Fallon cemetery. The remaining four of us—Velma Watkins, Lloyd Alcorn, Marie Ruppre, and this author, Ray Alcorn—still live active lives in Fallon, Nevada.

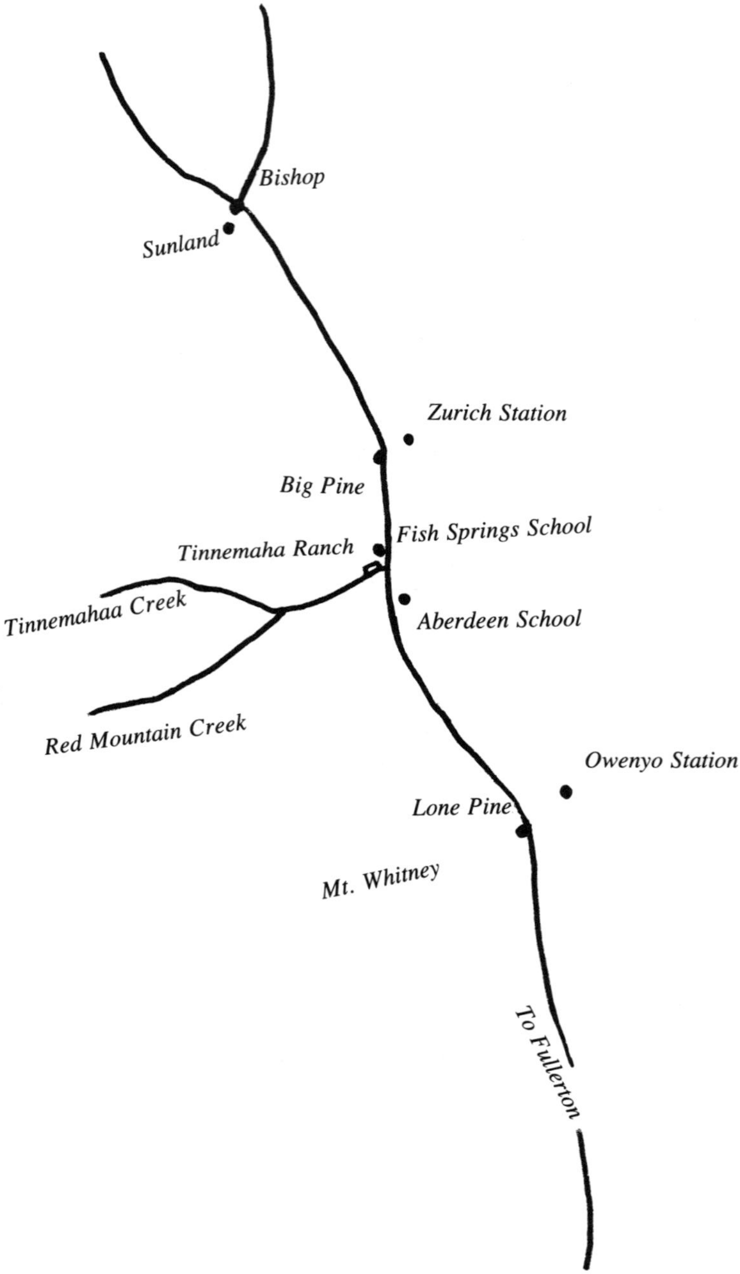
Bishop
Sunland
Zurich Station
Big Pine
Fish Springs School
Tinnemaha Ranch
Tinnemahaa Creek
Aberdeen School
Red Mountain Creek
Owenyo Station
Lone Pine
Mt. Whitney
To Fullerton

Chapter One
Westward: Kentucky to California

These are true stories of the early life of William Harrison Alcorn. His father, Charles Alfred Alcorn, died when he was a small boy. After his father's death, his mother, Nancy Elizabeth Branscomb Alcorn, left Kentucky with her children in a covered wagon to be with "family" in Kansas during the winter. Then they moved to southeastern Colorado, where they lived on a homestead. This chapter has stories that are both comical and tragic.

In the Monticello area of Kentucky, William Alcorn and his twin sister, Mattie, were born on February 20, 1874. Their parents, Charles Alfred Alcorn and Nancy Elizabeth Branscomb Alcorn, had nine children. They were Romulus Alfred, Emma, Mary Susan, William Harrison, Mattie, George W., Charles Wesley, and Linnie May. They had another child, Alice, who passed away at age two in 1880.

Charles, who was suffering from Tuberculosis, had been caught in a heavy squall while returning home from Monticello. He had to cross a stream which had swollen to a rushing torrent. It was necessary for him to ride his horse into the stream and swim to the other side. This severe exposure hastened his death in 1884.

Charles' widow, Nancy Elizabeth, left Kentucky in 1885, because the doctor told her to take the children to a dryer climate where they would not be so heavily exposed to germs. Traveling by covered wagon she took the children to Kansas where her late husband's brother, James Alcorn lived.

About 1886, Nancy Elizabeth's relatives decided to take government homesteads in Baca County, Colorado. Nancy was hopeful that the dry Colorado climate would be healthier for her children. She and her parents, Harrison and Polly Branscomb, left for Baca County, Colorado in a covered wagon caravan. The open prairie of Baca County was just over the line from

Kansas in the southeast corner of Colorado; the miles and miles of dry grassy plains were not like the wooded region that Nancy had left in Kentucky.

Life for the early pioneers was a constant struggle with many harsh and memorable moments. Living conditions were very primitive. They made sod or dirt dugouts for their houses. The Plain States—Kansas, Eastern Colorado, and Oklahoma—were and still are regions of extreme weather conditions. Severe winters with blizzards, and drought in the summer were not uncommon. These states were known for their frequent cyclones and most of the settlers had a cyclone cellar. What are known as tornadoes today were called cyclones by the early settlers.

Carroll Bryan (1966) related, "It took grit and courage and really tried a man's heart strings to leave his family in an unfinished shelter—sometimes simply a tent—while he went for supplies or to do some other necessary task. It took far more courage for the women to stay on the claim with one or more little ones while her husband was away from home. Her drinking water was probably in one or two barrels in the back yard. Often she went out on the prairie and picked up fuel. Yet stay she did, while at times the range cattle were too close for comfort. She knew the coyotes were really harmless, but their howls on those lonely nights were enough to chill the blood in the bravest."

* * *

After moving to Colorado Nancy Elizabeth married George Twyford, whom she had known when she lived in Kentucky. They had two daughters, Florence born in 1890 and Sophronia born in 1892.

Homesteading was a hard working, tough, isolated and lonesome life. Often there were many miles to the nearest neighbor. The settlers were without telephone or other means of communication except by walking, riding horses, or in a horse drawn wagon. They had a "close feeling" toward the land and prairie. Having a wife and children made life more pleasant; the family acquired a togetherness that was beyond belief. They all worked from early childhood, often from sunrise to sunset, not by choice, but by necessity. As a result, most of them by the age of fourteen were able to do all sorts of work without supervision.

But boys will be boys. One of my favorite stories, told by my father, William Harrison Alcorn, was when his stepfather, George Twyford, sent him and his brother out to hoe the weeds in the corn field. The boys decided since it was such a hot day, that they would go for a swim in the nearby river, to cool off, before hoeing the corn. Without realizing it, they played in the

George and Nancy (Alcorn) Twyford with their two daughters, Sophronia and Florence. Note the dog sitting on the cyclone cellar.

river most of the day until the sun was about to set. They quickly put on their clothes and ran for home, so they would not be late for supper. After supper they relaxed and as a full moon rose stepfather asked, "Did you boys get enough to eat?"

"Oh Yes," the boys replied.

"You're not tired from all that activity today, are you?"

"Oh no, we're O.K. . . ."

Then his stepfather added, "O.K., then lets go hoe that corn you didn't do today." They went to the field. A boy was put on each side of him in different rows. His stepfather said, "Now you just start hoeing and keep up with me." My father said they worked hard all night with their stepfather working with them. Father often wondered what would have happened if there hadn't been a full moon that night.

Most of the homesteaders had a feeling of dignity, self worth, and high morals. They were people who labored long hours for little compensation, but it was necessary in order to survive. As father grew up, he followed the customs of that period which was to work, work, work.

Father went to the hills to gather wood from the time he was a young lad until he left Baca County. Father once told a story about working for another man. The job was to haul wood from the hills. There were three wagons, each pulled by horses. The horses that father had been assigned were ''balky.'' A balky team does not take a steady pull. They have a tendency to jump forward and then rear back. The older men chuckled to themselves wondering how the ''kid'' would handle the horses. They didn't have any trouble getting to the wood area pulling an empty wagon, but getting home with a heavy load was another story. Father had worked with other balky horses and knew from earlier experiences, to stop on a down slope. After the wagon had been loaded, the horses ''hit the chains'' (harnesses) and recoiled back. The wagon was by then rolling and smacked the horses in the butt, forcing them forward. The horses pulled steady once the wagon was rolling. This took the skill of a good horseman. The next night, on the way home, father was lucky to find another down slope to park the wagon on. The process was repeated again the next morning, so there was no problem getting the wagon moving. As they traveled along, father heard to his delight, ''Boy, that kid is alright, he sure can handle that balky team.''

Bryan (1966) related, ''One of the necessities of life was wood, wood to burn and for posts to build fences. About sixty-five miles to the northwest in Baca County laid a region covered with cedar trees. The trips to the cedars were made during the summer and early fall. To make the necessary arrangements for the trips; one neighbor would contact the others and they would decide what each person should take and when to start. Those going on such trips were Edwin Elder and his brother Chris, Dave Konkel, Harry Alcorn, Jim and Jasper Konkel, also a man from the JJ Ranch on the Cimarron River and another man near Syracuse, Kansas joined us on the way. The last man had a ten horse team and was equipped to haul very heavy loads.

''Our last watering place was the Murray sheep ranch. We had to haul water in barrels to our camp sites. It was necessary that we have trail wagons—that is, two wagons coupled together. Most men drove four horses—teams in tandem. There were no graded roads nor any bridges. The roads were merely trails across the plains. In the bad spots it was necessary to uncouple the wagons and pull one at a time. After we were over the difficult places we would couple them back together and be on our way. By using this method we could haul much heavier loads of fence post and fire wood. Since the cedars were located on higher ground, it enabled the horses to pull very heavy loads.

''The first night out we camped on the plains northwest of Springfield near Two Butte Creek. To keep the horses from straying too far at night, as they grazed on the grass, we put hobbles on their lower front legs. When we removed the hobbles in the morning, the horses would head directly for the wagon to be fed their ration of corn or oats. No matter where the wagons were camped it was always 'home' to the horses. After we had cared for the horses and eaten our breakfast we would be off again for our next camp site. These trips took from seven to ten days, but aside from the hard work, they were usually a lot of fun.

"On one such trip, Uncle Dave and Harry Alcorn put up a small tent for their sleeping quarters. When all of the others were asleep Jasper and Jim Konkel had a great time pulling tent stakes, letting the tent fall. Neither Uncle Dave nor Harry said a word, but the next morning all were up at the break of dawn—much earlier than usual. This early rising continued throughout the trip until Jasper and Jim were far from any more 'nite pranks.' In fact, after the trip, Jasper was in bed for a couple of days recuperating. Jim did better, but was perfectly willing to sleep when given a chance.

"Fortunately for us, it seldom rained during the wood gathering trips. What rain did come was usually in the form of 'flash' rains and the water quickly ran off. If we did experience any continued drizzle that softened the ground we were in for a bad time with our heavy loads of wood."

William Harrison Alcorn in the center of the back row with his family.

Father told of a rainy trip when he became embarrassed beyond words. During the night he developed a bad case of diarrhea. It rained most of the night, so each time he got "the call," he went under the wagon to avoid getting wet. By morning he had covered the ground under the wagon. He saw the men chuckling and talking about the mess he had made. Finally, one of the men broke the ice and said, "What's the matter kid, you got a belly ache?"

Father replied, "I sure did during the night, but I feel better now."

In 1899, father was hired to drive a freight wagon. He told about one trip that he went on—the wind was blowing and it was very cold. He was afraid he was going to freeze to death, so he got off the wagon and walked alongside of the horses. The horses shielded him from the wind and gave off a little heat. The exercise stimulated his circulation, so he completed the trip without any problems.

* * *

This letter was written by Jim A. Alcorn who was Nancy Elizabeth's brother-in-law. The letter shows how religion played an important role in their everyday lives. Even though many people had very little schooling, a letter was always a welcome sight.

"Ionia Kans, May 2, 1890

Uncle Hary and Aunt

Will send yous few fines in Rom's letter. Nothin new at present. We are all tolarble well at present. Hope this will find yous all well. Children rote Aunt Polley was sick and had grip. I hope she is better by this time. It kep me in bed and bout house for weeks and as soon as I was up I went to Makato and my team ran away with me and threw me out and hurt one ankle so it laid me up for several weeks and it looked like I was havin my shear of ponishment. Hope Aunt is better now. Tell hur God Bless hur. If I never see hur here I want to live so I can meet hur on the other shore where we won't have pain and sickness and sorrows and when we leave this earth we can look down in our grave and say Lord sais it is enough, come up here. Uncle I hope you won't have grip. Do yous live by your selves or with someone. Has John ever come to see yous. Give all folks my respects and may God ever be your gard and Sheel al long as yous is permited to stay here on this trobelsom earth and may we all get hoome where we live for ever and ever. Rite.

Yours as ever J.A. Alcorn"

The Uncle Hary and Aunt mentioned in the letter above is Harrison

Branscomb and wife Polly, they were the parents of Nancy Elizabeth Alcorn. Polly Hamilton Branscomb died about a month after this letter was written.

* * *

In 1876, Simeon and Martha Ann Konkel moved their family in a covered wagon from Page County, Iowa to near Englevale, Crawford County, Kansas. This trip of about 200 miles took seven days.

Their daughter, Delpha Eululah, was born near Mulberry Grove, Kansas, on August 8th, 1881. She was five years old when the family moved by covered wagon to Baca County, Colorado.

Bryan (1966) reported, "The first frame house in Baca County built by a bonafide settler was that of Simeon Konkel, constructed in July of 1886. From that time on for more than a year the road past the Konkel homestead contained a continuous stream of prairie schooners—as the ordinary covered wagon was called. The Konkel home was the stopping place for many of the settlers.

"The first post office commissioned in Baca County, Colorado, was the one at Boston, which received its commission April 14, 1887, but that post office has long since been discontinued. Simeon Konkel was appointed postmaster of Stonington in 1894. It may be of interest to some people to know that the Stonington post office was commissioned January 20, 1888, and has been receiving and sending out mail since that time."

William Harrison Alcorn married Delpha Eululah Konkel on February 12th, 1900, at the farm home of her parents near Stonington, Baca County, Colorado. They had five children. Velma Merle, Lloyd Marvin, and Ruth Marie who were all born in Colorado. This author, Joseph Raymond, and Earl Jason were born later in California.

Father had plenty of outdoor work to do: plowing fields, planting, hoeing, and harvesting, taking care of the cattle, and hauling wood. Mother seldom worked in the fields. She cooked three meals a day, did household chores, and sometimes helped in the garden. She had a full time job washing clothes by hand with water carried in buckets from a well, washing the kids in a small tub, and mending the clothes that were often made by hand.

Inside plumbing was unheard of, the toilet, called an outhouse was a small building about fifty or more yards behind the house. There was usually a catalog in the outhouse to use for toilet paper, out in the open the people used grass. If it was night, raining, or snowing, a chamber pot was usually available.

Homesteading miles from other families, everyone learned to be self sup-

William Harrison Alcorn and Delpha Eululah (Konkel) Alcorn on their wedding day, February 12, 1900.

porting. Father and mother raised cows, chickens, and pigs. A vegetable garden and a few acres of corn were grown. The garden depended on the summer rains for moisture. Velma remembers eating potatoes that were peeled, sliced and fried. We had a lot of mashed potatoes, but Velma does not remember having baked potatoes. The family ate cooked dried beans an "awful lot." Corn was an important item, it was eaten as roasting ears in the summer and fall months. The family also ate corn bread all the time. Several scoops of soupy cooked beans over corn bread made a good meal. For breakfast, the biscuits were always made of white flour. Biscuits and gravy were an everyday occurrence. We made a syrup called Sorghum to be used as a sweetener on pancakes and other things.

Father and mother would take Velma and Lloyd to Vilas to pick up the mail and groceries once a month. It was a long trip, so the folks seldom went. They had very little money, so it was used sparingly. Most items purchased were of the utmost importance: flour, salt, and baking soda were the main items. The only time Velma recalls getting an orange was at Christmas.

Father always made sure there was a cow to milk. The cows produced plenty of milk when feed was available. The cows grazed on prairie grass. By late winter when the grass was short, the cows produced less milk.

To get cream, the milk was put into a large container and left overnight. By morning, the cream rose to the top and was skimmed off and saved to make butter or to use for desserts. To make butter, the cream had to be whipped or agitated by churning. Most settlers had a wooden churn with wooden paddles to plunge up and down in a constant motion until the cream separated into butter and buttermilk.

These churns were used by the Alcorn family to make butter. One is a crock churn and the other is a wooden churn, both have wooden paddles.

Mother canned a lot of food that was used in the winter. Eggs were processed in the summer, by pouring a solution of sodium silicate called ''Waterglass'' over the eggs to preserve them for later use.

Some corn was saved for seed to plant in the spring. In the fall, the corn stalks were harvested and used to feed the cattle. The chickens were fed scraps from the table and in the summer they ate grasshoppers and scratched for bugs. In the winter they were given limited amounts of corn. At night the chickens roosted in the chicken house. At daylight they foraged for food. Hawks, coyotes, and foxes were a problem. When a hawk captured a chicken some member of the family would rush to scare the hawk away. If the chicken was wounded or killed, it was cleaned and cooked for dinner.

The family occasionally went to sunday school and church in Stonington. The long journey each way prevented frequent trips. On one trip, my brother, Lloyd remembers that the preacher was invited for dinner. They killed a rooster to give the preacher an appropriate Sunday dinner. Any guests were welcomed whether it was family, neighbors, or strangers.

Velma recalls that when she was a little girl, some people came by in a wagon and stopped to talk. Lloyd, her little brother, wanted to show his strength to the visitors, so he grabbed her and threw her to the ground. She was really embarrassed!

* * *

One of the various problems that our family, as well as other families in the area had to deal with, were range fires. As the grass became dry, there was always a danger of fire, which was often caused by lightening. All the people were prepared to fight prairie fires with buckets and barrels of water. A wet gunny sack was used to beat the fire. With little or no wind to fan the fire this method was very effective. Sometimes children playing with matches would start a fire; it could quickly spread to become a very serious problem.

Bryan (1966) reported that George Ervin Konkel learned from the older boys how to play with fire in tall grass. ''They would let the grass burn a little then whip it out. Ervin thought he would try this at home, which he did a number of times quite successfully. But one day when Pa and Ma were gone and his older brother Willie was in the field, not realizing the wind was too high for such stunts, he tried this game once too often! The fire got out of control and burned about a quarter mile of rail fence belonging to a neighbor, and a nice young orchard of his Uncle, Jerry Konkel, before the neighbors could get the fire under control.''

Our family did not escape these fires either. Velma and Lloyd were playing with fire which caused some problems. They were playing in the barn that Velma said was made of "soap weed," when they decided to build a fire under a calf muzzle. These muzzles were made of wire and temporarily tied over the nose and mouth of the calf, so the calf could not nurse the cow. Velma was using one of the muzzles as a play stove. As time passed by, one of them looked up and the wall of the barn was on fire! They both ran to the house and reported the fire to mother. Father was away at the time, having gone to the hills for wood. The neighbors saw the smoke and responded with barrels of water. But it was too late! The barn was burned to the ground. There was one setting hen in the barn that also perished in the blaze. Velma still remembers the husky voice of a neighbor when he said to mother, "Well, did you give them a good lickin'?"

Weather was another item that gave cause for alarm due to the cyclones and lightning. My father told of his brother, Rom, being killed by lightning. He sadly said, "It was a direct hit and killed him instantly. It tore his boots off!" Father always told us, "Sometimes it is best to lie down if you get caught out in a thunder storm, because standing up you are like a lightning rod."

Cyclones were something everyone was aware of and talked about. When a storm was approaching, everyone would watch to see whether it developed into a cyclone. When a person was caught out in the open, they would try to find a low place and lie down, so they would not be carried away. Father told me, "At one time during a cyclone, a man grabbed a hold of a small tree to keep from being sucked away. After the cyclone passed he was completely covered with mud."

Velma remembers, when she was a little girl about eight years old, traveling with father, mother, Marie and Lloyd to visit father's twin sister, Mattie, and her family who lived in Oklahoma. The trip took two days and on the way a storm approached. They stopped the covered wagon at a nearby ranch to seek shelter. The rancher said they could take shelter in the barn. As they went into the barn for cover they saw the rancher and his family going into their cyclone cellar. Mother thought that those ranchers were an "awful lot" not to let them go into the cellar for such a short period of time. Inside the barn, mother and father carried on quite a conversation about whether they should stay in the manger, in the corner of the barn, or go out into the open and find a low place to lie down before the cyclone struck. Fortunately, the cyclone did not hit, so everyone was happy, but mother and father's conversation was remembered by Velma and Lloyd all their lives.

* * *

The prairie was open to all farmers to graze their cattle. Lloyd remembers when dad spotted some cattle heading for his unfenced corn field. Father quickly saddled up a horse and sent Lloyd out to drive the cattle back. Lloyd said, ''Those cattle were determined to eat that corn, I had a heck of a time turning them away.'' Sometimes fencing was used only on small tracts such as garden areas, but in many cases fencing supplies were not available.

Cattle drives were made each year from far away places to where the summer feed was more sufficient to fatten the cattle. Some cattle drives went from Texas to Wyoming. On these drives, the cattle were rounded up for the night and would bed down till morning. Sometimes the cowboys were heard singing in the evenings before bedtime. Father, on occasions, would ride to their camp to visit and sing with them. The following day the cattle would be moved out at an early hour. When some of the cattle became sick or weak for various reasons and could not keep up with the rest, they were abandoned. They were called ''dropouts.'' Father, by nursing some of the dropouts back to health, was able to acquire a few cows at a time. Eventually father had forty head of cattle.

Velma remembers as a young girl, she and her brother, Lloyd, were given the daily chore of picking up cow chips (dung) from father's cattle, which ranged near the house. They put the chips into a gunny sack and when the sack became to heavy to pick up, they would drag it to the house to be used for burning in the stove.

After the cowboys left camp and moved the cattle on, there were many cow chips left on the bedding ground. Our family waited until they dried before they went out to pick them up. Father would then hitch up the wagon, and he and the kids would gather up the chips and haul them home.

On one occasion, there was a cattle drive camp within a mile of home. After the cowboys moved out, looking across the prairie, father saw a wagon several miles away heading straight for the abandoned camp. Our family badly needed chips for the stove, knowing very well that these people would take all of the precious chips, father sprung into action. He harnessed the horses, hooked up the wagon, picked up the kids, and went galloping out to the camp area. They found that the chips weren't quite dry, but who cares when fuel is short! They managed to get their share. On the way home father commented, ''We got our share of that oozy stuff all over us too.''

* * *

It was a constant struggle to keep the animals protected from the weather, the coyotes, and other predators. Earlier stories reported many wolves and buffalo. The wolves disappeared after the buffalo were killed off. One time a wolf had killed one of father's cattle, so he stopped everything to set traps near the carcass to capture the animal. Sure enough, the very next morning a wolf was in the trap. That was the last wolf that father ever saw.

While father was in Baca County, the coyotes were abundant. Frequently, coyotes were seen very early in the morning. Father had told us about shooting his .22 rifle at a coyote that was near the chickens. The coyote ran away with such speed, that father stood there in awe. He was sure he had hit that coyote. While watching the coyote as it ran, about four hundred yards away, he saw it fall. After retrieving the animal he found the bullet had hit the coyote's heart.

Father told an interesting story about a coyote he saw as he was riding his horse across the prairie. The coyote was hiding behind a clump of grass. Father was about one hundred yards away, when he started circling around the coyote. The coyote kept hiding behind the grass as father made two complete circles. Then, father galloped his horse straight for that coyote and immediately the coyote took off running.

Some coyotes hid in old buffalo wallows. The wallows were made years earlier by buffalo pawing the ground to get the dust and dirt over them to protect themselves from insects. The more a wallow was used the bigger and deeper it became. Some wallows were about six feet across and about one or two feet deep. They also were used as watering holes after a rain.

* * *

There were many rattlesnakes in the area and everyone was aware of their presence. One of our cousins, who was about two and a half years of age, had been playing and running behind the older children in the yard, when she was bitten by a rattlesnake and died. The children were always advised to walk in the middle of the road to avoid any snakes that were hidden in the grass at the edge of the road.

Rattlesnakes were often found in corn fields, so care was always taken when walking in these fields. Lloyd remembers when father heard a snake rattling and sent him out into the open. Father rushed around stomping the Kangaroo Rat holes closed, so the snakes couldn't crawl into them and hide. Father then searched the area and killed the snake.

Some people were frightened by the rattlesnakes presence, especially in the corn fields where a snake could easily hide. Corn was an important crop and family members and neighbors often assisted in hoeing and cultivating the crop. Father and two other men were hoeing the corn, when one of the men grabbed his leg and cried out, "Quick men, a snake!" They ran to the man's side to lend assistance. He was bent over and had a grip like a vise on his leg just below his knee. Father asked, "Did it bite you?"

"I don't think so," the neighbor replied.

Father commented, "Well I don't see any snake, where is the snake?"

The neighbor replied, "He ran up inside my pant leg and I got a hold of him, I'm afraid to let go, he'll bite me." They felt around but could not find the snake.

Father said, "I don't see how it could be a snake."

The neighbor replied, "Well, I got him clamped against my leg and I can feel him." They finally got the man to release his grip and out dropped a kangaroo rat. Father commented on this incident years later, "That poor ol' kangaroo rat was dead, dead! He had the life squeezed out of him."

Velma recalls, "Uncle Wesley dug a pit to put snakes in. It was about six feet across and there must have been a hundred rattlesnakes in there. Of course I was only a little girl when I saw them, so the pit may not have been as deep or wide as I remember it and there may have been fewer snakes. I never could figure out why Uncle Wesley put those snakes in the pit."

Uncle Wesley was the prankster and happy-go-lucky brother of the group. The brothers thought of a plan, to have some fun with Wesley. This was to have him steal some of father's chickens as a prank. However, unknown to Wesley they told father what was to take place. They told him to be sure to leave the gate open because, when Wesley became alarmed, he would certainly run the shortest route out of there, and they didn't want him to crash into the gate and break it. At the appointed evening father concealed himself in the chicken house and in due course, in the darkness, he could see Wesley approaching. He entered the chicken house and father fired the gun. Sure enough, Wesley turned and ran with great speed out of the area, father laughed for days. They told that story for many years afterwards.

* * *

During severe weather the family seldom had visitors. People only traveled when it was necessary, because winter blizzards were an unforeseen problem. A person could loose his bearings and become lost as visibility became difficult in the blowing snow. To keep from becoming lost, people on the farms often had a wire or a rope to guide them, which was tied from the house to the barn and outhouse.

During blizzards, cattle usually turned their tail into the storm, so they would not have to face the wind. Sometimes they would slowly drift to a low wash or gully. The snow would pile higher and higher filling the gully, until the animal was unable to get out. It was important to find these stranded cattle and pull them out before they died.

Winters were difficult for the cattle and by spring many of them were skinny and weak. Once they laid down in this condition they were to weak to get up and eat, so they would die if left lying on the ground. Father kept a close watch on them and would "tail up" any of them that were down. The "tail up" process was done by getting hold of the tail and twisting upward on the tail. This seemed to give the animal extra strength to get on its feet. I suspect that it was a painful experience for the cattle to be tailed up, but it was necessary to save their lives.

* * *

As time went by, it became necessary to send the children to school. Velma was nine years old and Lloyd was seven. Neither had been to school because there were no schools nearby. Father decided to sell out and move further west. The family thought the change might also help father's sinus and allergy problems.

Before leaving they hitched up the wagon and visited various friends and relatives in Oklahoma. On one visit, our relatives put up a swing in a big tree, so the children would have something to keep them occupied. They also played with their gentle old mule. The next morning, to their horror, they saw the mule with it's neck stretched up in the swing rope and it was sort of sitting down. Rushing out to rescue the mule, the men soon realized it was dead. The mule had gotten the rope twisted around its neck and hung itself, which was very sad.

Our family left Baca County and moved to Greeley, Colorado where Velma and Lloyd attended school. Velma said she was very shy during her first few days. She told the teacher that she never got to use any of the swings. The teacher responded, "You have to be more aggressive, when recess is out you have to run out there and grab the swing first." This was not what she had been taught as a child on the prairie.

The next year the family moved to Lebanon County, north of Durango, Colorado. Father cleared trees from the land to cultivate it. Father had a lot of sinus trouble while they were in Durango. Dad's sinuses were so bad that they thought he might get Tuberculosis; he constantly used a salt solution to alleviate the congestion in his nose. It was fascinating for Velma to watch him. The first time she saw him use the salt solution, she thought dad was only going to drink it, but instead he closed his mouth and forced the solution out his nose.

Mother's brother, Will Konkel, who lived in Fresno, wrote to mother about California and thought it would be an excellent place for them to acquire farm land. Since, our family was not satisfied with where they were living, because of father's sinuses they decided to go to California.

Within a year the family boarded a train for Fresno. Our family took only a few necessary personal possessions and a box full of food for the trip, so as not to rely on the train dining car. Lo and behold, the train soon became snowbound for three days and the dining car ran out of food. Everyone was complaining about being hungry, so our family shared their food.

Arriving in Fresno they were greeted by mother's brother, Will Konkel, who helped them find a house nearby to rent. Velma and Lloyd went to school there for about six months. Father thought of settling there, but land prices were "sky high," so he went to New Pine Creek, Oregon. There he met a Mr. Tichenor who was also looking for a place to settle. They heard of a place called Owens Valley, California, so the Alcorns and the Tichenors went there. They liked the valley and the nearby mountains so much, that they rented a large house in Bishop where they took their families to live.

Chapter Two
Newcomers—Sunland

This is an account of the Alcorn family moving to the Sunland area and how they acquired their land and made a living. The ranch in Sunland was about three miles south of Bishop. The nearest school was about one mile away. This is a series of short historical stories about the Alcorn family, while they lived in Sunland.

The Alcorn family arrived in Bishop in the fall of 1911, just in time for Velma and Lloyd to start school. Marie was not yet of school age.

The Alcorns and Tichenors rented a large house, that was sectioned off, with the Tichenors living in one part and the Alcorns in another. Father and E.D. Tichenor were compatible people and enjoyed each others company. They thought it would be nice to buy adjoining or nearby ranch property, so they could trade work back and forth. This was a custom often practiced in the early days when one needed help. You work for me and I'll work for you attitude.

An issue of the *Inyo Register* of Bishop/Inyo County, California dated Thursday, February 9, 1911, had a heading "BIG DEVELOPMENT ENTERPRISES." This article states, "First payments of purchases totaling upward of $200,000.00 will be made on over 2000 acres of Sunland District lands within one week. The tracts will be thoroughly improved and marketed by a million dollar company." This article is an example of how the land was sold in large parcels and then broken up and sold into smaller tracts. Father and Mr. Tichenor were considering buying some land south of Bishop where they could plant orchards and irrigate the land. They also talked about starting a nursery and furnishing small fruit trees to the valley. It was these smaller tracts that father and Mr. Tichenor decided to purchase.

In the local Bishop/Inyo newspaper, *Inyo Register,* dated November 30, 1911, the heading read "ESTABLISHED NURSERY." It reported, "Sunland acreage purchased for the purpose of establishing a nursery. Sale of 20 acres of land by P.W. Forbes to E.D. Tichenor and 20 acres by A.J. Allison to W.H. Alcorn were made this week in Bishop Land Company. The price paid was $5,600.00. The tracts are directly south of the main street in Sunland district. The investment is primarily to start a nursery to supply the city and the increasing demand in Owens Valley for good fruit trees. Apples will be given principle attention to growing of all trees likely to be in demand on part of the business. These gentlemen are in earnest about this venture and will go ahead at once with their arrangements, as their purchase is unbroken land so far, their first plantings will be made on leased ground. The advantage of a home nursery goes beyond the mere fact that its trees are acclimated. It avoids the risk of injury in long distance shipments and assures a large percentage of successful transplanting. The enterprise is of merit and timeliness.''

* * *

While the Tichenors and Alcorns were sharing the big house, two children were born. A son, Joseph Raymond, this author, was born to the Alcorns on

Delpha E. Alcorn got caught weighing herself.

December 1, 1911. I joked that my birth rocked the earth, because ten days after I was born they had an earthquake. I do not remember any of this, but it is fun to think that, maybe, I shook some of those high Sierra Mountains. Mother must have been shaken up too, because in the family bible she had written 1912 as the date of my birth. Later, she crossed out 1912 and wrote 1911. Mother's doctor did not record my birth. Because of that, I do not have a birth certificate.

Velma says she remembers the earthquake of December 11, 1911. They told everyone to go outside including Mrs. Tichenor, who was in labor at that time. After the earthquake was over, everybody went back into the house; a short time later, Mrs. Tichenor gave birth to a daughter that they named Hazel. For a while, we thought Hazel would be born outside.

Moving to Sunland in the spring of 1912, the Alcorns felt like they were finally settled for a lifetime. Sunland District was a little farming community. It should be noted that main street was a far cry from a city street. It was a dirt road with three houses scattered over a mile stretch of road. The Bantas owned land north of Maine Street in the Sunland District, not far from the Alcorns.

The twenty acres father bought, had a small two story house on it. There was no running water in the house. For our water supply we built a cistern by digging a hole in the yard to a depth of about eight feet by four feet in diameter. The bottom and sides were then cemented to prevent the escape of water. This hole was covered with boards to prevent us kids and animals from falling into it. The cistern was filled by diverting the water from an irrigation ditch. Then it was filtered through a box of charcoal to purify the water as it entered the cistern.

There wasn't any electricity in the house so they used kerosene lamps. All meals were cooked on a wood burning stove. The wood stoves were also used for heating. The outhouse (toilet) was a small building built in the back yard, away from the house. The outhouse had two holes, however, some outhouses had as many as three or four holes for a large family. If there were more than one hole, one often was made small for the children. Old Sears catalogues usually ended up in the outhouse, for toilet paper. Colored pages were used last, because they were stiff and slick, not easy to use. There was a moon crescent cut in the door that was for ventilation and identification.

The irrigation ditch that ran near our house usually had water running through it. Early in my life I had my first swim in this irrigation ditch or should I say my first unintentional swimming lesson. The irrigation ditch ran under a bridge and on the other side of the bridge was a little water fall that dropped about two or three feet. My first lesson was brief. Marie put me in a tub and sent me under the bridge and over the falls. The tub tipped over, I landed in the water and began screaming bloody murder until Marie pulled me out.

Father was a good natured person and he occasionally played jokes on us kids. His jokes were always in good taste. One time at a very young age, I was in the orchard when a rock fell from the sky. I knew Chicken Little would believe me, but no one else would. The sky couldn't be falling! Later I figured that dad had tossed the rock, even though he never said anything about it to me.

Velma was given money to purchase items in Bishop. She had gotten a gold coin back for change, at that time gold coins were common. It was a five

William H. Alcorn ready for work.

dollar gold piece, which was a lot of money in those days. As she was walking home she dropped it in the sand. She looked and looked without finding it. She rushed home to report her loss. Father and mother returned with her to the sandy spot in the road, but they never were able to find the gold coin.

Velma does not remember father working for other people. Most of his time was spent working in the nursery and orchard. Lloyd remembers one time going with father to a ranch near Manzanar where they planted pear trees.

During fathers spare time, he went fishing and hunting. He frequently went up Rock Creek to fish. When the fishing was good and he caught more fish than the family could eat fresh, he would put them in a barrel and salt them. As he added more fish, father would add more salt and layer the fish in the barrel. Several months later, when we were ready to eat them, a few fish at a time were taken out and soaked overnight in fresh water to remove the salt. Most of the fish that the family ate were fried. The only time we would bake fish was when the fish weighed four or five pounds.

Marie and Ray with their father, Harry Alcorn, showing off their trout.

Father and mother always stressed the importance of our schooling. Even when we were out of school, we discussed various subjects. The Sunland School, as well as many other small district schools, was built because of the lack of transportation.

In the *Inyo Register* of September 5, 1912, an item under Sunland Department stated, "School opened Monday the attendance was 35 with Mrs. M. A. Clark in charge." The attendance of 35 students was considered very high for those days. Mrs. M. A. Clark was remembered as being a husky lady, but very nice and never having any problems as a teacher. Sunland School was about a mile and a half west of our house. It was within walking distance.

Bill Banta and Lloyd went to Sunland school together and became good friends. All of the children that went to school took their lunch and usually ate together. Velma and Marie shared their lunch. Velma, being the oldest, divided up the lunch. Marie always accused Velma of giving her the worst part. Velma would give Marie a sandwich and in the presence of the other school kids Marie would say, "Is this all you got, an old piece of dried bread?" Velma would become so embarrassed that she wanted to crawl in a hole.

Velma remembers, "Us girls always had to wear dresses to school. The dresses went down to the ankle or mid calf. I would get tired of wearing my long underwear because it would show. Well, I found a cure for that! As soon as I was outside of the house, I would pull my long underwear above my knees and go to school, then on the way home, I would pull my long underwear down before I would go in the house. It would have to be awfully hot before our parents would let us girls take our long underwear off."

Lloyd said he remembers when Earl was born, September 23, 1915, "Mother was in labor, I had to take the buggy and bring Mrs. Tichenor back to help. But, before we got back, Earl had already been born."

I was not old enough to attend school, but sometimes I went with my sister as a visitor. In order to leave the room at the Sunland School and go to the outhouse, you had to raise your hand, and get the teachers permission. I needed to go, so I raised my hand. At the same time another kid held up his hand. The teacher excused the other kid first and by the time he got back, I really had to go. I made a mad dash to the outhouse, but it was too late. I cleaned up the best I could and went back to class. Later the teacher was standing near me and said, "I smell something." She sure did, but I was not going to say a word. As soon as I got home mother said, "You smell like you had an accident." Needless to say, I got an extra bath that week. At that period of time baths were taken in a round light metal tub, on every Saturday, whether we wanted it or not.

Ray Alcorn, the author, as a small child being left behind, while the older children went to school.

The high school in Bishop was located about three miles from the farm. Father had an old Model "T" Ford car that was used for transportation. When Velma was a freshman in high school she finally got the chance to drive the Ford. She was taught how to crank it to get the car started, how to drive and change a flat tire.

When Velma first started to drive, she had some problems. The automobile tires of that era were narrow and one could easily get stuck in the mud or sand. Several times, on a muddy road when the car was swishing around, she wondered if she was going to survive the experience. On one occasion she parked near the outhouse. When she started the car, she put it into the wrong gear. Instead of backing up, she went forward and pushed the outhouse off its footing.

Velma used the car as a school bus. Not only did she drive Lloyd, but she also drove two neighbor kids, the Edwards, who lived up by Rossi Hill from our place. Velma was about 17 months older than Lloyd. Our parents did not permit Lloyd to drive because they thought he would "act up."

When Lloyd went to high school in Bishop, he would meet his friends at the pool hall during lunch time and shoot some pool. Lloyd said he was allotted twenty-five cents for lunch, he could buy a coca-cola for five cents. The pool hall was the only place that the boys had to "hang out."

The pool hall was located on the corner of Line and Main Street in Bishop. Mr. E.R. Moser, at one time, owned the pool hall. He had two boys, Glen and Jim. Lloyd said, "We would play pool and bet a quarter, if you lost, you went without lunch that day, consequently, I soon became a good player."

* * *

Not every newcomer appreciated the life in California, like our family did. In 1914, Uncle Eldridge and Aunt Bessie Bryan moved their family to Sunland from their farm in Colorado. They rented a house near the Sunland School. Uncle Eldridge worked for neighboring farms, with the thought that he might soon buy one. After two years they decided they could do better in Colorado, so they moved back to their original farm.

The Walter Shober family enjoyed the Sunland area like our family. They lived about half a mile from the Sunland School. The Shobers did general farming, such as raising alfalfa and wheat. After they sold their farm property, they stayed in the valley and bought a place they named Shober's Resort.

In talking with some "old timers" from Owens Valley, they all reported they had enjoyed living there. Mr. Harry Corkill in Fallon (1991) said, "My life in Owens Valley was very good."

Mr. Corkill came to Bishop in 1913, at the age of fourteen from the Isle of Man. He immediately started working full time for Mr. Danny Karmode. He was paid ten dollars a month, room and board, and they also washed his clothes. He milked, by hand, about thirty cows twice a day and did other related work. At times he watched other young people going to school and was saddened, because he had to work and could not join them. Later he worked for Frank M. Andrews who had thirty-five milking cows. As he grew older he was paid thirty dollars a month working for George Waterson.

A man told Mr. Waterson that Harry Corkill was a good man and Mr. Waterson said, "No, he is a good boy." Later in life, Mr. Corkill acquired ranch land, but sold it to the city of Los Angeles and left the valley in 1927, with no regrets.

Chapter Three
Tinnemaha—The Ranch

This chapter is about the lives of the Alcorns, as well as others, who lived on the Tinnemaha Ranch from 1915 to 1929. These true stories were taken from this authors notes, memories and information given to me in recent years, by my brother, Lloyd, sisters, Velma and Marie, and from various friends.

The Tinnemaha Ranch was located in Owens Valley, California about nine miles south of Big Pine. "The name Tinnemaha came from a Piute legend that the great Medicine Man of the Piutes was Winnedumah, brother of Tinnemaha, War Chief of his people" (Chalfant, 1933). The creek, lake, mountain, and reservoir presently carry the name "Tinemaha." The new spelling with only one "n" is used on recent maps of the area. While we were living there, it was spelled with two "n's." The Tinnemaha Ranch, at various times, was called the Red Mountain Fruit Ranch, the Burdick Ranch, the Chapman Ranch, and sometimes locally referred to as the Alcorn Ranch. The Alcorns lived there from 1915 until the ranch was sold by Charles C. Chapman to the City of Los Angeles. Our family left the valley in 1929.

The Tinnemaha Creek that flows through the ranch begins high in the Sierras merging with the Red Mountain Creek above the ranch. Much of this water was used to irrigate the orchards and fields.

The Red Mountain, a dormant volcano, that adjoined the ranch is covered with red cinders, hence the name. Poverty Hill, which was between the ranch house and the main Owens Valley, ascends at a steep angle to the east of the house. Looking toward the west, about six miles, one could see the Sierras. We were surrounded by mountains and hills.

Father started working on the Tinnemaha Ranch in the early spring of 1914. Over the years, he had planted all the fruit trees on this ranch. Vincent Burdick, at one time, owned the ranch with other people until he lost controlling interest. They left Burdick as manager, until they sold it to Charles C.

The Tinnemaha Ranch before the big house was built.

The Tinnemaha Ranch with Red Mountain to the left. Note, at the bottom left, the stone pillars which marked the entrance to the ranch.

Chapman in 1917. Mr. Chapman gave my father complete charge of the ranch which included signing the payroll checks. Father was offered wages, groceries, and expenses for the ranch. Always thinking of his family, father thought it would be a good permanent place to live.

Lloyd remembers, "The Tinnemaha Ranch had a total of 2700 acres: about two hundred acres of apple trees, forty acres of pear trees, and eighty acres of alfalfa and pasture land." The remaining acreage was open range. The shade trees were mostly Locust and some Lombardi Poplar. Father also planted several acres of Locust trees southwest of the hill orchard, which were used for fence posts, because they were more durable.

Even though Mr. Chapman, the owner of the Tinnemaha Ranch, paid the grocery bill, we were always conservative with our purchases. Father and mother set down rules as to what would be bought. Every Saturday, they would go to town with a beautiful matched team of horses named Rock and Rye. Later Mr. Chapman, bought a 1917 Dodge four door Sedan from Hazard Garage in Bishop for father to use. The car was used for trips to town to get supplies and various trips to Mr. Chapman's office or residence in Fullerton, which was about 300 miles or a days drive from the ranch.

The Tinnemaha Ranch with what the Alcorn family called the Hill Orchard at the far left.

The large Tinnemaha ranch house with Poverty Hill in the background.

Mr. Chapman seldom came to the ranch, but when he did us kids looked forward to seeing him. He would always have a silver dollar in the palm of his hand, which he transferred into our hand with a handshake.

For the first couple of years the house and the bunkhouse were in the same building. This old house had a good sized living room, a couple of bedrooms, and a dining room that had a long serving table. A rope hung down from the ceiling over the middle of the table. This rope was used to ring the dinner bell that was attached to the top of the house. The kitchen was a large square room, with a huge wood stove. This stove was large enough to put a boiler on to heat water on wash day, and still have room to cook the meals. The bunkhouse, where some of the workers stayed, was located on the south side of the house.

Later, Mr. Chapman built a new large ranch house for us to live in. It had an ample sized kitchen where the family spent much of their time. The living room, on the east side of the house, was used mostly for company. Us kids were told to stay out of the living room. The upstairs had two bedrooms, plus a big hall like room. Our family held several dances upstairs, as well as having roller skating parties on various weekends. Velma said, "I learned

Ray Alcorn standing in front of the Tinnemaha Ranch house.

how to roller skate upstairs." One time somebody fell down and broke their leg or ankle and that ended the roller skating. Marie remembers, "I was not allowed to go upstairs when they had dances, because I was too young." I do not remember my dad having dances upstairs after the roller skating accident.

The Tinnemaha Creek flowed within a few yards of the ranch house. There were a lot of rocks in the bottom of the creek, that we called cobblestones. They had a smooth surface and were of various sizes. These rocks were used in 1920 to build a stone front porch which was the finishing touch to the new ranch house. From the front porch of the house a person could cast a fish line into the creek. Us kids caught many trout out of this creek.

The road entrance to the ranch had two huge stone pillars that were made of the larger stones. Each pillar was eight feet high by three feet in diameter. Father put a single tree in each pillar for historic heritage. These pillars still stand today. When I was a young boy, I hid my arrowheads under one corner of a pillar, but when we left in 1929, I dug them out.

* * *

Earl Alcorn on right, with relatives, kneeling near the water wheel which generated electricity on the Tinnemaha Ranch.

In order to have electricity a wooden flume was built to carry water at an elevated height. By opening a gate, the falling water would turn the water wheel which was connected to a dynamo that generated electricity. The water wheel was turned on every evening at dusk and turned off at bedtime. Water was also piped from the head of this flume into the house, with a fall of about fifteen feet, giving us sufficient water pressure.

Father decided to change the course of the Malone Spring that was located on the lower slope of the Tinnemaha Mountain, because it ran down the hill through the sagebrush and seeped into the soil or evaporated. They wanted to divert the water to the Red Mountain Creek which was a short distance away. There was about thirty inches of water coming out of the spring. We had no surveyors to tell us where to dig the ditch. In order to get the perfect flow, we used hand shovels to make a small trench from the spring to start the water around the hillside. A small amount of water was diverted into a ten foot long trench, which was extended a few feet at a time. By using the water as a level, the proper fall was easily determined. It had to have enough fall, so the water would run through the ditch. However, too much fall and the water would wash a deep channel and eventually wash the hillside away. Dad told Lloyd to get two high school kids that wanted to work. He got George Green and Albert Ray. They set up a camp at the work site during Christmas Vacation. Dad told Lloyd, "Be sure and tell the boys not to let the water run at night in the new ditch, because it will wash out the hillside." About two days later, Lloyd took some food up to the boy's camp. When he arrived, there was a big gully in the mountain. He said, "What in the world happened?" They told him, "Well, we forgot and let the water run at night." Boy did it wash a big gully, which is still there to this day.

In an effort to conserve water, a concrete ditch was built several miles above the ranch. As I remember it was about a half mile long, eighteen inches deep, and fourteen inches wide on the bottom. The water ran swift, because the ditch was steep and sloping. If an animal fell in, it was not apt to get out. Coyotes, snakes, and rabbits were some of the animals caught in this ditch. One day father was taking a drink from the end of the concrete ditch, when he looked up and saw a dead rabbit. Father later asked Dr. Menning, a family friend who was visiting the ranch, what could happen if he drank water with something dead in it. Dr. Menning said, "It would not be half as bad as a person, at the head of the creek using it for a toilet. By the water running down the creek, it would purify itself, before it got to the ranch." Father then felt better about drinking the water. Father always asked Dr. Menning about any health questions we had.

Father hired men to do the irrigating, but Lloyd also helped. We had to irrigate about twenty acres of apple trees on the hill orchard. Father remarked how well Lloyd could irrigate the hill orchard. The hired men would get too much or not enough water through the rows, but Lloyd could adjust the water perfectly.

A four or five inch metal pipe that ran down the hill was used to run water into concrete take-out boxes. Every spring we had to clean out those metal pipes. The pipes were about ten feet long. If we could not see light at the other end, we knew there was something in the pipe. We would grab the pipe and turn it straight up in the air. The cottontail or whatever was in the pipe would fall down to where we could reach it. If it was a cottontail, we would grab it and have it for dinner. The rabbits never attempted to bite us when we grabbed them. One time, I turned a pipe straight up, to find out what was in it. I started pulling out a bunch of sticks and trash. I thought, ''Good gosh, what is in there anyway?'' About that time I felt something jump around, I grabbed it, and pulled it out. Lo and behold, it turned out to be a wood rat that had made a nest in the pipe. I quickly let go, and away he went. It is a wonder he didn't bite me.

There were a lot of people hired to work on the ranch. The sheepherders and camp tenders were mostly Basque. They spoke very little or no English. As many as fifty Indians were hired during apple harvest time. The small colony of Indians that lived permanently on the ranch lived in five wickiups made of willows. Father hired an Indian by the name of Bob Harry, as foreman of the Indians. Bob Harry met with father every morning to discuss what was to be done. Men were hired for pruning trees, as well as other odd jobs. Father paid some of the hired help forty dollars a month plus room and board.

At the Tinnemaha Ranch most people addressed my father as Mr. Alcorn, very few called him ''Harry.'' He was a man with high principles; he was honest, hard working, very proud, and friendly to others. I do not ever remember him arguing with anyone. If father had to scold someone or advise them, he would talk in a moderate manner. Therefore one would never feel belittled or shamed.

During prohibition time, there was no booze allowed on the ranch or at the Indian settlement. Some people still managed to bootleg whiskey. Father never took a drink himself. Two men came to see my father, one of them was a politician. He got out of the car and shook father's hand, displaying a bottle he said, ''Here, Mr. Alcorn, I want to buy you a drink.'' I can see father to this day very tight lipped saying, ''I never touch the stuff.'' The other man said, ''Maybe if we hold him and pour it down him, he might like it.'' Father told them, ''If you try, you will get a few lumps on your heads.'' My father had his own convictions. I idolized him for standing up for those convictions. For that very reason, I went for a long time in my life without ever tasting whiskey.

* * *

Ray and Marie herding the yard flock of sheep on the Tinnemaha Ranch.

When we first moved to the ranch, there was a dairy herd of about 130 Holsteins cows. The diary cows were on the ranch from the time we moved there until 1920. In later years, we had over 100 white faced Hereford cows and 10,000 sheep.

Years before Mr. Chapman acquired the main herd of sheep, a small yard flock of about fifty sheep were on the ranch. Dad used to have Marie and I herd the sheep to the pasture, which was about a quarter of a mile from the house. There was not a fence between the pasture and alfalfa field. Marie and I did not dare let the sheep get into the alfalfa or they would bloat and die. All Marie and I had to do was stand there and keep the sheep out of the alfalfa. Boy, I sure did not care for herding those sheep.

When we acquired more sheep, they were in flocks of about 2,000 with a herder looking over them. Lloyd was camp tender one summer. He went on horseback with two pack mules, taking food and supplies to the sheepherders. Since Lloyd did not speak Basque and they did not speak English, the herders would go through his pack and pick out what they wanted.

Father checked the sheepherders periodically to see how they were getting along. Sometimes from the ranch at night we could see the herders campfire, as far as six miles up the mountain. Father would sight over a post, facing in the direction of the campfire. He would wiggle his feet around to make a good imprint. The next morning, standing in his foot prints and sighting over the post, he could tell approximately where the sheepherders were camped. One time, a ranch-hand went with father to check on the herd. He asked father, "How did you know where to find the camp?" Father told him about the post and footprints. At the ranch there were many practical things that a person had to do, that was our way of life.

Since the sheepherders and camp tenders spoke little or no English, father had trouble communicating with them. One time my dad said to a herder, "Now, you are going to take the sheep past that man's ranch down there. I do not want you to go over into his field at all. Stay a quarter of a mile away from his place." The sheepherder said, "Oh, just go in a little bit, just a little bit, in that green field. The sheep really like that green stuff." My dad was disgusted with that sheepherder.

One time, father let a camp tender take the car to town to get what supplies the sheepherders needed. He had trouble driving a stick shift. It took him about two hours to go to town and two hours to come back, even though it was only nine miles. He put the car in low gear and drove to town and back going about four miles an hour. It took him forever to get the supplies back to the ranch. On one occasion another camp tender went to town to contact a sheepherder. When he got home, he left a note for father, that said, "I not could find no sheepherder no how." It made father laugh.

Some of the sheepherders were excellent at handling the sheep and keeping coyotes away. The sheep were only kept on the range in the summertime. In the Fall, we drove them to the area around Bakersfield for the winter months.

Men that father hired to work on the ranch were from various backgrounds. These are some of the men I remember. A single fellow by the name of Hummel stayed in the bunkhouse and worked at the ranch during 1917. Another was an English draft dodger during World War I. Then there was Picken-Paugh from Southern California. Picken-Paugh had only one or two teeth, which us kids thought made him look strange, but he had a great sense of humor. Then there was a horse trainer, about the size of a jockey. All he talked about was racehorses and tracks. His wife helped mother cook while they were there. Fred Lockhurst, a single man who had a beautiful voice, played a small accordion and harmonica. They were all good workers.

Delpha Alcorn on the left and one of the ladies that helped her cook.

Father had a great voice. When everyone was singing, such as during a social event, father would join in. Many of the social events were held at various schools. I recall him singing to the family at times. My gracious, could he yodel! I can remember as a small boy my father singing outdoors in a yodeling fashion, "Fe Fi Fo Fum, I smell the blood of a Englishman, Hoo Wee, I'll get cha, I'll get cha, Hoo Wee, I'll get cha." Then, if we were in the right place, the echo would come back from the hills in a softer voice as if it came from a stranger, "Fe Fi Fo Fum, I smell the blood of a Englishman, Hoo Wee, I'll get cha, I'll get cha, Hoo Wee, I'll get cha."

We seldom had visitors, normally when someone stopped at the ranch it was either for some information or road directions. I think the most outstanding "character" that ever came to the ranch was a one night visitor. As evening approached one of us kids said, "There's someone coming on a horse." Now this got everybody's attention and they craned their necks to see who it was.

Mother asked, "Well, who is it?"

Father commented, "I don't know, I've never seen that horse before." A little later he said, "Oh my gosh, it looks like a woman."

Now this really got everybody's attention. It could be said that any stranger approaching created a stir among the family and hired hands, but a woman was unbelievable.

Father and mother met the lady rider. She said, "I'm traveling through the country. I'm hungry and need something to eat and a place to sleep for the night."

During that time, a woman did not travel through the country alone. They invited her to come in and told her she could stay the night. Father took her horse to the corral, while mother showed the lady her room and prepared some supper for her. In the morning the lady departed after an early breakfast, to heaven knows where. After she left, father and mother carried on quite a conversation.

Father: "My, but that was a stinking woman."

Mother: "You should smell the room she slept in."

Father: "I never smelled anything like that before."

Mother: "It was a funny smell, it smelled more like a disease-dirty smell."

Father: "How are you gonna clean that bed? Should we burn those blankets."

Mother: "No, I'll wash them, but first I'll boil them for a half hour." Mother washed and boiled the bedding. She also washed the bed and the wall next to the bed. Then she let the room "air out" for quite awhile. After that, my parents both agreed that they had better not tell the next occupant about the lady that had slept there.

Another unexpected guest came to our house one night. At that time there were a lot of men working on the concrete ditch which was being built above the Pear Orchard. Marie and Velma were sleeping upstairs across the hall from Lloyd. Velma woke up and heard someone walking in the bedroom. It was as dark as it could be, Velma could not see anything, but she knew it was footsteps because she could definitely hear them. She shook Marie and woke her up. As Marie woke up, she raised her arm and hit someone. Whoever it was, ran out the door. It was a windy night, so they never did hear him go downstairs. They woke the folks up, who were very concerned. They thought it might be Lloyd prowling around or one of the workers from up the hill. They went into Lloyd's room and woke him up. Lloyd said, "I haven't been in the girl's room." From then on, our family started locking the doors at night which we had never done before.

On occasions, Marie and Velma went into the bunkhouse and talked to the men. Soon our parents forbade them to go, because it was not proper. The only exception was when the girls were accompanied by father and mother in the evenings to hear one of the men play his accordion.

When Marie was about eight years old, the girls were forbidden to go into the corn patch. My parents were concerned because Marie played a game with Picken-Paugh, where she would run down a row of corn and hide, until she was caught by Picken-Paugh; he would then rub his whiskers on her face. Father and mother said that it was not proper to be chased by men in the corn patch.

Mother's teachings were impressive. Cleanliness, honesty, and hard work were stressed in our family. We were required to take our bath on Saturday evening, whether we wanted to or not. She always repeated, "There is no excuse for being dirty. Poor, yes, but not dirty." We also were taught, kids should be seen and not heard. When anyone arrived at the ranch, father met them in the driveway, I often followed behind. Father would do all the talking and I only listened to the conversation.

We were trained to submit to control and yet be able to think for ourselves. Father told us boys, "Never hit girls or women, treat them with respect." Most men in the 1920's wore hats and when they passed a woman on the street, they would tip their hats. Some boys wore caps, that had ear flaps folded inside which they pulled down in cold weather to cover their ears.

Mother remarked, "Manners are important because a person's reputation is very important in the community." I recall father saying, "If you see a woman smoking a cigarette, you can just figure she is no good! Absolutely no good." He told my sister, "Don't ever smoke a cigarette, because everybody is going to think you are no good and that you are one of those street women."

Father and mother loved us very much. They showed us tender care and tried to keep us from getting accustomed to self indulgence. Father believed that everybody should work and have a certain amount of discipline. We were also taught to accept responsibilities, like when father would tell me to take the horses to the base of the mountains. I would be expected to saddle the horses and ride to the designated spot and meet him there, no questions asked.

Father once said to me, "Mother wants us to get a Christmas tree." Father and I went on horseback toward Stickers Bench to find a tree. We came to an isolated tree that was near the foot of the mountain. Father said, "Why don't we cut three limbs from this tree, it will not hurt the tree a bit. These limbs will make a christmas tree, and we will not have to ride all the way up the mountain tiring the horses." Returning home, mother exclaimed, "Oh, you're back already, let me see the tree." Father showed her the limbs and she asked, "Is that the tree?" I will never forget the expression on mother's face. After father tied the limbs together and put it on a stand it looked great and mother was happy.

* * *

Velma on the cobblestone porch, at the Tinnemaha Ranch, waiting for her date.

Velma remembers, "A boy came to visit me. While we were sitting on the couch talking, we heard something from above. Looking up we saw Marie's leg dangling from the ceiling. Marie knowing that I was getting all the attention, had gone upstairs and removed the cover from the stove pipe opening. This opening was used in winter months for the temporary stove. She put one leg into the hole and was swinging her leg around. I was so embarrassed! I called mother to take care of Marie and her antics from the attic."

According to mother, as a young fellow I had a terrible temper. I can still hear mother saying, "Son, you've got to learn to control your temper." Frequently Earl and I would wrestle around on the ground. Earl was a spunky kid and very often he would disagree with me. Even though there were some disagreements, I had great love and respect for all of my brothers and sisters. Families in those days had a greater sense of togetherness and wanting to help one another.

Earl and Ray Alcorn having lots of fun on the Tinnemaha Ranch.

It was customary for the older brothers or sisters to help with the young children. My younger brother, Earl, and I received good training from our sisters. They were a significant influence on our learning of manners. If we were chewing with our mouths open or trying to talk with a mouthful, they would really let us hear about it. Those teachings instilled in me a great respect for my sisters.

Father and mother never argued in front of us. Mother taught us that married couples should air their differences away from their children. Mother was a sweet lady who kept us very much in line. The threat of father in the background kept us lined up even better. We knew very well any punishment dished out by father would be more severe.

Earl once told me in confidence, "If mother whacks you with the razor strap for anything, cry out loud—'ouch, ouch, that hurts,' then she won't give you a second whack." Prior to my brothers advise, I would take three or four whacks without saying a thing. Taking my brothers advise, I would holler after the first whack and not get another. You would think, I could have figured that out for myself.

Velma and Marie sitting in rocking chairs taken from the house to have their picture taken.

Velma said, "As a young girl I was talking to some people and used the word darn. After the people left, I was scolded by mother for using such a bad word."

It amazes me now, when I think back, how well we were disciplined. We were all taught how to act when we were in someone's home; we were to sit in a chair, and keep quiet. We also were not to enter into the adults conversation, unless asked to do so.

Father also taught us the proper way to handle a gun. We were told all guns were considered loaded. The most important lesson he instilled in us was, "Never point a gun at anybody whether it is loaded or not." As we grew up this kind of discipline and training made for a law and order society.

Everyday, Marie and Velma would help clean house, cook, and do other household chores. Lloyd did everything, from mowing the hay to picking apples to irrigating. I do not remember doing everyday chores. Father paid me five cents for each pocket gopher that I caught or trapped. I helped make apple boxes for two to five cents a piece and later took care of the bees. Marie

Lloyd and a hired hand putting up the hay.

said, "The only time I got paid from Mr. Chapman was when I packed apples." Velma said, "When I cooked for the men, I was paid the regular cooks pay." Lloyd was able to make enough money to buy a car in 1918 by getting a regular paycheck from Mr. Chapman. The car was a 1914 second hand Ford and it cost him four hundred dollars.

Mother cooked the meals for the hired hands. Marie and Velma helped along with other women such as Emma Harry, Mrs. Blake, and other hired men's wives. Baking bread was a daily chore, because buying a loaf of bread was unheard of.

Our bread was made of white flour and baked in a wood stove. The pans for making bread, would hold six to eight loaves. Velma said, "I used a pan for six loaves, because that was all I could handle. My arms would get tired from kneading the large amount of dough."

One day when Marie was making bread and putting the dough in the bread pans, Bill Banta, a friend of the family, told her she was filling the pans to full. Later, Marie looked into the oven and sure enough, the bread was all over everything.

One summer, Velma worked in the cook house to earn money for a wrist watch that she wanted. She got up at four o'clock every morning, went to the cook house, and cooked breakfast. We usually had ham and eggs, or bacon and eggs with biscuits. Velma got to the point that she could make biscuits as good as mother's or Aunt Ethel Bryan's.

One morning after the folks left to do business in Big Pine, Velma wanted to have dinner ready for them when they got home. She decided to make spaghetti. She was not sure how much to cook; she ended up making about ten times more then what was needed. Mother wanted to know why in the world she made so much spaghetti. All along Velma thought she was doing something great.

Father did not like cabbage. One time, mother cooked some cabbage and camouflaged it with herbs. Dad ate it and did not say a word. Mother said in confidence to me after dinner, "You know, dad ate that cabbage and he didn't even know he was eating it." She was so pleased about it. An hour later, I was out helping father and he said, "Boy, that cabbage sure gave me indigestion. It tasted terrible. Mother is always trying to feed me cabbage." Father used to say, "The man worthwhile, is the man that can smile, when his wife feeds him cabbage for dinner."

I was not like father, because it did not matter to me what I ate as long as there was a lot of it. I could really eat and was always hungry. In the Fall, I constantly nibbled on apples. I could eat about a dozen to fifteen apples a day.

One time father and mother were going to be gone for the day, so they left me some money to buy what I wanted. I thought I could really enjoy myself; so after school I went to Josephs General Store in Big Pine and bought a dozen rolls and a quart of mayonnaise. When I got home I ate the whole quart of mayonnaise on the dozen rolls; for a long time after that I could not bear the sight of mayonnaise. I did not get sick, but I had enough mayonnaise to last me a long time.

Dr. Menning had told my dad, "Eating a raw egg will give you quick energy. It is a lot better for you than a piece of candy." Dad said to me, "Did you ever eat a raw egg?"

I answered, "No."

He said, "I will show you." He cracked an egg, tipped his head back, swallowed, and down it went. I tried it, boy, that egg sure didn't feel very good when it went into my mouth. I hardly swallowed and it slid down so fast I hardly knew what happened. I did not relish the idea of eating another raw egg.

Some of our hired hands had a funny way of eating too. Mother passed the strawberry jam around to use on the biscuits. One gentleman set the jam on his plate and started eating it. I guess he thought it was some sort of dessert, so he ate it all.

Another time a similar thing happened, a hired hand sat down to eat. The olives were in a little dish next to him. He did not realize that the olives were to be passed, instead he put them in the middle of his plate and ate all of them. Everyone had a different opinion about whether or not he was dumb or smart, because he got to eat all of the olives.

There was always a big garden and in the Fall, mother canned a lot of fruit and vegetables. There was a rock cellar on the ranch where we put fruit and vegetables for the winter storage.

We had a large watermelon patch, from which we sold melons in the late summer. When the watermelons were ripe, we would pick them and throw them up to a man in the truck. When one of the men missed catching a melon, everyone would sit down and eat the broken watermelon. I suspect a few were broken on purpose.

The apple trees were sprayed twice a year with lime sulfur or arsenate of lead. Once when the petals of the blossoms were falling off and again after the apples had started to grow. We had two sprayers, each pulled by a team of horses.

During apple harvest time, to pick and box the fruit for transportation, it took about fifty extra people, mostly Indians. Both men and women worked.

Ray and Marie irrigating the garden at the Tinnemaha Ranch.

We picked the pears green and put them into smaller boxes. This was the only way the pears could be shipped without bruising. The apples were sorted in various sizes, then each apple was wrapped in tissue paper before they were boxed. The majority of the apples were of the "Old Winesap" variety.

The boxed fruit was taken to the narrow gauge railroad at Zurich near Big Pine, about twelve miles from the ranch. Lloyd said, "There were as many as 27 narrow gauge car loads of apples in one year." When the narrow gauge met the broad gauge railroad in Owenyo, near Lone Pine, the fruit was transferred to the broad gauge railroad for the final trip to Los Angeles. The broad gauge having larger cars, took half as many car loads as the narrow gauge. When the apples arrived in Los Angeles, they were put in refrigerated storage until sold.

* * *

Ray and Earl shelling corn for the turkeys and chickens.

Every year we raised and prepared meat for the ranch. We had pigs, turkeys, chickens, and when deer season opened, we had venison. The way we used to kill turkeys was to hang them up by their feet and put a metal weight on their lower beaks. With a thin sharp knife, we would cut the blood vein in the roof of their mouth and pierced the brain. This caused the relaxation of the muscles which allowed the feathers to be plucked easier.

Our family butchered the hogs when they weighed approximately 250 pounds. We butchered in the late fall, because the meat kept better in cooler weather. There were several ways of killing the hogs. We shot them with a .22 rifle in the forehead to stun them, then immediately stuck a knife in their throat, so they would "bleed out." As soon as the hog was bled out it was scalded.

The process for scalding a hog was an interesting thing in itself. A big iron kettle in the backyard held about fifty gallons of water, which was heated over an open fire. In order to get the right temperature, we dipped our finger in the water three times. If we couldn't get our finger into the water a third time, the water was the perfect temperature. The hog was dipped in the hot water a few

Lloyd, Ray, Earl and Marie on the milk cart which was used to transport the milk from the barn to the separator house.

seconds, to loosen the hair. We then tested to see if the hair was coming off by using a scraper. If the hair would scrape off, we then pulled him out and put the other end in the hot water to repeat the process. When the scalding was done, the hog was put on a bench so the hair could be scraped off. This process took less than half an hour. The hog was then gutted and hung up usually over night to cool.

The fat was trimmed from the hog, cut into small pieces, placed in an iron vessel and heated. This fat would melt into a liquid. This was called rendering the lard. The heated liquid lard was then poured into small containers such as empty coffee cans and left to cool. We used the lard in pies and for frying and cooking. The solids that were left in the bottom of the kettle were called cracklings. Us kids liked to eat them. Mother would use the cracklings in cornbread and other foods for flavoring.

When the hog was cut up, some meat was used right away. The bacon and hams were salted and combined with other spices to keep all winter. Sometimes hams were smoked to preserve them. Some fresh meat was fried, put into a big stone crock, and hot lard poured over it. The meat would keep during the winter months using this method.

The lard sometimes was left in the vats to make soap. This was done outside in a big cast iron kettle. We would add lye to the liquid lard, as well as other ingredients, when cooled, it was cut into squares to be used for soap. It was used mostly for washing clothes. Emma, Bob Harry's wife, helped mother on laundry day, the clothes were washed in a round tub, with a wash board. This lye soap was too strong to use on a person's body.

* * *

To milk the cows we would herd them from the pasture into the milk barn. The milk was put into ten gallon cans then put on a push cart. This push cart was on a rail track that went one hundred yards from the milk barn to the separator house. The separator house was by the creek, so there was plenty of running water. Before the milk was separated, we took out what fresh milk we needed to drink and cook with. (The separator is a device that separates the cream from the milk.) The cream was put into five or ten gallon cans and sold to the creamery. The skim milk was usually fed to the hogs and the cats, who eagerly awaited their share.

In the big barn, I remember watching the horses in their stalls eating corn. Boy, listening to those horses chewing that corn was sure noisy. Located in the barn was a harness room. Also in this same area was the blacksmith shop, where Lloyd learned how to make hammers and other things from Bob Harry.

Near the harness room there were old pieces of equipment, such as wagons and an old buggy, which we used before we got the car. The buggy had leather upholstery, where some wasps had built their nest. I had been playing in the buggy, when the wasps started stinging me all over the face. Bob Harry was doing some work nearby and to this day, I can still hear him yelling, "Get out of there, run! Run! Get out of there!" The wasps stung me numerous times, causing my eyes to be almost swollen shut. After that painful experience I always remember, when you get stung, start running.

* * *

Young boys could get in a lot of trouble when they weren't under supervision. One afternoon I rode my bike to Jimmy Dowd's house when his folks were not home. We decided to ride the calves. Jimmy held the first struggling calf, I threw my leg over his back and clasped my arms around his neck. The next thing I knew I was bucked off and in the dirt, laying flat on my back. Jimmy was laughing at me, acting like my ride and fall was one of the funniest things he had ever seen. When he finally recovered from his laughter, it was his turn to ride the calf. His ride was over as quickly as mine, but from where I was standing it wasn't funny. Jimmy hit his head, got up, and acted sort of dazed, so I helped him into the house. After he was safely in the house, I closed the door, jumped on my bike, and peddled home. It was some time before I saw Jimmy again, but I was glad to see that he was alright.

Us kids found many ways to keep ourselves amused. Some of the games we played were marbles and mumble peg. Mumble peg was played with a pocket knife on a soft board or a lawn. We flipped the knife and it was suppose to stick a certain way in the soft board. The knife position determined wether we would receive twenty-five, fifty, seventy-five, or one hundred points.

We made things to entertain ourselves. We used a quarter inch thick piece of wood, which was two inches wide and about a foot long. We cut a small hole in the end of the board where we tied a string that was two or three feet long. Whirling this board around our heads it would make a weird zoom, zoom noise.

As a small boy, I played outside in the dirt. I made roads, hayfields, haystacks, and all sorts of things. I built myself a ranch in the shade of a tree. Mother's friend said after seeing my spread, "Well, that boy is going to become an inventor when he grows up." Whatever gave her that idea I do not know, but mother repeated this to me many times after that.

When I wasn't on a horse, I was riding my bicycle. I remember going through the gate at the same time my dad was breaking a horse to ride. There were trees all around, so neither one of us saw the other coming. When the horse saw me on my bike, he started bucking. Dad and the horse went out through a plowed field, dust was flying so bad, I could not tell what was going on. The next thing I knew, the horse came running out of a cloud of dust, with no saddle or dad. I thought, "Oh my gosh, maybe the horse killed dad." I was really scared. When the dust finally settled, I saw father sitting in the dirt. He said, "The cinch broke, I flew off, and came down, still in the saddle. It was a good thing that it was soft dirt or I may have broken a bone."

One day, mother and Aunt Olive were riding their horses. Mother said, "Here, have an apple." Well, she tossed the apple, thinking Aunt Olive would catch it, but she missed. The apple hit Aunt Olive right on the end of her nose and gave her a nosebleed. It is a good thing mother did not hit the horse on the end of his nose or Aunt Olive may have ended up on the ground.

Lloyd was a good horseback rider. He was the one who often broke the horses, so he expected them to buck. I was never on a horse that bucked with me.

Father and I took the car to meet some deer hunters as they rode off the hill. As I drove up, they stopped to open the gate and one of the horses started bucking and bucked a man off into the sagebrush. I was determined to ride the horse that had been bucking. Catching the horse, I rode it back, leading the other horses to the corral. He did not try to buck with me. The hunters rode back with my father in the car. They had been riding all day and the horse should have bucked at the beginning of the day instead of the end of the day.

Some horses never became used to approaching cars. A car coming down the road, would scare a team of horses, causing them to run away. An Indian was working with a team of horses on the ranch, when a car passed by and scared the horses. They ran away, causing the man to fall into the farm equipment injuring himself.

Going to Bakersfield with father, we had to go over a mountain pass before we got to the valley. In the 1920's, almost everyone had an automobile, buggies were not used much anymore for transportation. Father said, "I want to show you something interesting." Father showed me a dead horse that was still in its harness, hooked up to a buggy. It had been there many years. There was nothing left, but a dried hide. I wish I had taken a picture of it, but I don't recall having a camera at that young age.

* * *

We always had dogs, but they never came into the house. There was so much activity around the ranch, I do not think they wanted to come in. We had a dog that chased cars. Dad said to me, ''Boy, that dog is going to get killed. Anytime a dog starts chasing cars, he will soon get run over, especially if the cars go fast.'' Most of the time the dog ran alongside the car biting at the wheels. Dad said, ''I'll have to break that dog of doing that. He's such a good dog, I don't want him to get hurt. I've tried lots of ways to get him to stop chasing cars. I'm going to have to be pretty severe. Ray drive the car and I will get in the back. When he comes out to chase the car, I will throw a rock at him.'' Dad picked up a rock, gee whiz, it was two or three times larger than the size of a baseball. He said, ''I have to find something to soften this up.'' Father got a burlap sack and tied it around the rock. He told me, ''I don't want to hurt the dog, I just want to knock him down, so that he will know not to chase cars.'' The dog did not chase us when we left, but when we started back, out came the dog. Father hauled off and threw the rock at him. Luckily, it hit him broadside and knocked him down. The dog let out a yip and headed for the house. Sure enough, he never chased a car again.

Many of the fences around the Tinnemaha Ranch were wire netting. Above the three foot netting fence there was a strand or two of barbed wire. Someone suggested dad get some greyhounds, because they would be great for catching and keeping the jackrabbits under control. Someone loaned a pair of greyhounds to him. One day, dad started to go through the barbed wire fence and one of the greyhounds jumped at the same time, both got caught in the fence. The dog started to bite dad and everything else around him. Needless to say father did not crawl through any fences if the dogs were nearby.

* * *

The Indians lived pretty much to their custom. The women wove baskets out of willows. They dug taboose grass which had an eatable nut on the root; it was plentiful in the valley. The women pounded these nuts into a meal and added enough water to make a paste. Lloyd said, ''They put a blanket in the middle of the tent and placed the bowl of taboose meal in the center of the blanket. Everyone sat around the bowl and with two fingers they ate out of it.''

There were not many pinenuts in the nearby area, but the Indians gathered them from the White Mountains to the east. Going to the Indian colony settlement with an Indian boy, we went inside a wickiup and saw a hundred pound cloth flour sack that was full of shelled pinenuts. The Indian boy took a little handful of the shelled pinenuts and offered them to me. Boy, they sure were good.

These hand woven baskets were made on the Tinnemaha Ranch by Emma Harry's mother, Mattie, in 1922–1923.

Father and the Indians had a field day one Sunday. There was bow and arrow shooting, horseshoe pitching, sack races, tug of war, and things of that sort. The competition was high, but no one was able to really out do the other, except when the Indians showed up with their old bows made of native wood and their bow strings made of sinews. My father, having a modern hickory wood longbow, out shot all of them.

One of Bob Harry's relatives passed away. It was the custom at that time to burn all of the departed's belongings. After the belongings were burned, Bob Harry went to see if everything burned completely. As he was walking around, he kicked a can, and something rattled inside. He picked it up and to his amazement, there were a few gold coins in the can. Bob sold them to my father. Father had them for many years, until the government went off the gold standard and asked everyone to turn their gold coins over to them. Father turned them in to the bank and was paid face value.

When I was 10 or 12 years old, I went with Bob Harry on horseback on different occasions to move cattle. I usually carried a slingshot around my neck and a pocket full of roundish rocks. I saw a rattlesnake not far from us and shot it. Bob Harry said, "Don't kill him, he'll bite you."

I said, "Well, he can't bite me. He's already dead."

He replied, "His relatives will bite you."

I asked, "How would his relative know who killed him?"

He assured me, "Oh, the sagebrush will tell them." Some people have different beliefs and superstitions. Bob Harry had told me many of his beliefs, that I can no longer recall.

* * *

The lane that ran for a mile straight up toward the Sierras from our house was a dividing line for hunters. Father had a rule that anybody who hunted quail could only hunt on the north side of the road. By enforcing this rule, he kept the south side of the road for a reserve. Father only let the people who asked, hunt on the ranch.

Bob Harry told father he had shot some quail. Father asked, "Did you shoot them on the wing?"

Bob replied, "Oh, I shot them in the wings, feet, tails, all over." To shoot them "on the wing" means you shot them while they are flying. That was quite a story, that was told for many years.

Father had another rule that nobody could hunt in the orchard, but he had an awful time enforcing it. Some people did not follow father's rules. As a consequence, father asked me to go with future hunters and show them where to hunt. I went hunting several times with Mr. Hepler, who was a wealthy man from Southern California. I took Mr. Hepler quail and deer hunting. We did a lot of deer hunting on Stickers Bench.

My folks usually invited Mr. Hepler to stay for supper. Mr. Hepler was quite a talker. Glen Moser, Velma's husband, thought, "By gosh, I am not going to let him dominate the conversation during supper." Mr. Hepler being a religious man, father asked him, "Will you say grace?" Glen did not realize Mr. Hepler was saying grace, so Glen started talking in a real loud voice to cut Mr. Hepler off. Glen raised his voice and Mr. Hepler started raising his voice. Velma, sitting next to Glen, began poking him in the leg, trying to get him to be quiet. He asked, "Why are you poking me?" Then, all of a sudden, he realized what was going on and became quiet, real quick. Later, Glen told everyone, "One of the most embarrassing moments of my life was when I was talking during grace."

After some years went by Mr. Hepler gave me a 16 gauge double barrel Ithaca Field Grade shotgun. I got a lot of use out of that shotgun. At first I

Bob Harry, the Indian Supervisor, is shown here with his deer.

could not hit anything, but I soon learned to aim accurately. Several times I shot at flying quail, eight or ten shots, without missing one.

It was interesting to go hunting with the men from the city. It was like taking someone horseback riding, who did not know which side of the horse to get on. Some city dudes were not knowledgeable about the ways of the outdoors. These city people seemed to have their guns always pointing at somebody. Father enforced the rule—never point a gun at anything, unless you intend to shoot it.

One time, when I was about fifteen or sixteen years old, I took some horses to the base of the mountains where some dudes from the city were

Harry Alcorn and Glen Moser with three nice buck deer.

going to hunt. They were to eat their breakfast at the ranch and drive up to the base of the mountains to meet me at daylight. Since it was six miles to the mountains, I started out before daylight. I rode a couple of miles up the road when I saw something in the middle of the narrow dirt road. The horses shied away from it. All of a sudden a guy raised up from his sleeping bag with a rifle in his hand. I did not realize that there were any other hunters in the area. I said to him, "Good Morning, I am on my way up the mountain." I thought it was strange that he was sleeping in the middle of the road. Another person was with him, but I only focused my attention on the guy with the rifle. Their car was parked where it could not be seen. We always figured if a person slept or parked their car in the middle of the road, they were Los Angeles dudes. In other words those people were not experts at being out in the country. On the other hand, those dudes probably thought we were a bunch of hicks, that did not know anything about city life. Had we gone to Los Angeles, we probably would have done all sorts of oddball things.

* * *

The Alcorn Family. Back row left to right, Marie, Harry, Delpha, Velma and Lloyd. Front row left to right, Earl and Ray.

Lloyd recalls, "The biggest snow storm I remember while living at the Tinnemaha Ranch, was around Thanksgiving in 1917. One morning I woke up and heard the folks talking about a big snow storm. When I looked out the window, I was surprised to see so much snow; it must have been over two and a half to three feet deep."

It took two hours to shovel a path approximately one hundred yards to the barn. The snow was up to Lloyd's waist. Thirty-two dairy cows and stock horses had to be fed. Before taking the hay wagon out to the field to feed the cattle, we had to tromp the snow down.

Lloyd recalls, "By the end of the second day of the big snow storm, there was a total of four or more feet. Father saddled Cap, the big horse, to go up to the upper pasture and check on the other horses. Cap took four leaps, squatted down, and would not move. We had sleds and skis that were made by Bob Harry and some of the ranch workers. Father went to the pasture on skies and found the horses eating willows, so he knew they would be okay. Thirty days later, finding them all alive they were able to bring the horses to the corrals."

We were stranded at the ranch about a month. Fortunately, we had plenty of groceries and supplies, but we weren't sure about our neighbors. When we finally got out, dad and a couple of men went almost a mile down the hill on horseback to check on our neighbors, the Conners. After the snow thawed, we were able to take the horses to break a trail, in order to get to Big Pine. The snow was so heavy it caused the roof of the hall, which was used for dances in Big Pine, to cave in.

Snow caused problems with traveling. We could not use the buckboard, because the snow was too deep, so we rode the horses to school. Marie and Miss Shumaker rode old Buck and Lloyd rode his own horse.

During the big snow storm, one of the Indians on the ranch passed away. They had to bury him in the snow until the ground thawed enough to dig a grave. While he was in the snow, the other Indians danced every night to keep the spirits away. Lloyd recalls, ''Velma and I sneaked out in the dark and watched them dance. This dancing continued for about a month until the ground thawed enough for the man to be properly buried.''

With the towering peaks overshadowing the valley, like a giant mysterious spirit, it is no wonder the earlier Piutes developed certain beliefs regarding the area.

We had our share of windstorms at the ranch. One of which was practically a hurricane. It broke the upstairs window, near the bed where Lloyd was sleeping.

Coming from Southern California, father and I encountered a strong wind storm north of Mojave. We put rocks in the back of the car, so the wind would not blow us off of the road. The dust was thick, father advised me to keep driving, so we would get through the dust storm. Within a mile, we were through the worst part. On arriving home, we saw that the dirt, fine gravel, and sand, had nicked the paint on the car. The wind blew so hard, that some gravel made small holes in the isinglass curtains. The curtains on the old cars had a plastic window called isinglass that let the light through, so a person was able to see out the window.

* * *

During World War I, a flu epidemic caused some anxiety with our family. No one would go anywhere, not even to school, for fear of getting the flu. The folks did not even go into town for groceries. Lloyd said, ''Bishop High School was closed down for two months. I think thirty-eight people died from the flue in Bishop.'' Since the Tinnemaha Ranch was isolated, nobody contacted the flu. We still kept in touch with others by way of telephone.

When I was young, the glands in my neck became swollen. We were not sure if I was developing a goiter. The doctors put iodine on my throat and made me sunbathe. My parents put me in a wheelbarrow and had Velma make sure I stayed in the sun for the time required. The cure must have worked, because I felt better. An old wives tale was thought to be, if a person with a goiter smelled the breath of a horse, it would cure him.

I was bitten by a strange dog and everyone was afraid I would get rabies. Luckily the dog did not puncture my skin, so I did not have to go through the treatment. This treatment consisted of receiving shots for over a week on alternating sides of your midsection.

Our family was always getting slivers. We would run to our mother and she would take them out. When I was about six years old, I was walking around

Ray when he was about six years old.

the bunkhouse and ran a sliver in my foot. Mother got some of the sliver out, but not all of it. For a long time my foot was sore, finally it got so bad, my folks took me to a doctor at Big Pine, I think his name was Dr. Crook. He gave the folks some medicine to put on my foot. After a couple of weeks, father said, "I am going to take that boy to Bishop to Dr. Shute, because his foot is not healing." Father took me to Dr. J. W. Shute's office. His office was on the east side of the street, upstairs overlooking Main Street. He had me lie down on a table, and said to dad, "Now, hang onto his leg, so that it will not move." Dad had one hand on my chest and one hand on my leg. Without giving me anything to deaden the pain, the doctor took his scalpel and cut right into my foot. He removed a inch and a quarter long sliver. I screamed so loud that it was probably heard all over Bishop. Dad was real amazed, by the size of the sliver that was removed from my foot. Not long after the sliver was removed, my foot got better.

* * *

Approaching Big Pine from the south, on the left hand side of the street, was Kispert's Garage. On the right was Mindenhall's Photo Shop. Mr. Mindenhall took pictures and his wife ran the gift shop and painted. Josephs Store was about a block north from Mindenhall's.

When I was very young father seldom took me to town, because I would get car sick. He told me a story about one of the few times, he did take me to town. On our arrival, I got out of the car and the first thing I did was let out a big holler. Father thought, "Gee whiz, what is that for." About two blocks away another kid let out a holler. Father later said, "I got a kick out of you and the other kid. You would think that you kids had been hollering to locate each other all of your lives. Pretty soon, the kid came along to where you were and you said to the kid, 'Do you want to fight, kid, kid?' " Dad calmed me down. I figured I had to fight, because someone told me that the town kids wanted to fight me.

* * *

During my last year of school in Big Pine, I drove an old Dodge flat bed truck which was used for the school bus. The truck had seats running down both sides and a top over it. I never drove the bus more then about fifteen to twenty-five miles an hour. The kids got in from the back on makeshift steps. Driving along, an Indian boy, by the name of Tiffany Hill, would jump out

Big Pine 8th grade class of 1926. Our teacher was Zella Hudson Joseph. Back Row left to right, John Coe, Mack McDonald, Floyd Ketchum, Ray Alcorn and Tiffany Hill. Front Row left to right, George Cortnel, Virginia Rossi, Marie Haugh, Zelma Carnel and Edward Blake.

of the bus and run along behind it. He would jump back on the step, until he was rested, then he would jump down and run behind the bus again.

* * *

We had about 100 bee hives on the ranch. They were placed around the orchards in bee yards to help pollinate the fruit blossoms. Each yard contained 20 or more bee hives. They were placed in rows a few feet apart. With all of the apple and sweet clover blossoms, the bees made excellent light honey. The honey we removed from the bee hives was called comb honey, because the honey was not extracted from the wax.

We removed the honey from the bee hives before the wild Buckwheat bush blossomed, because the bees produced a strong dark colored unpalatable honey from this plant. This honey was left for the bees for the winter.

One day when I was attending the bees and looking in the top of a bee hive box, I felt something on my foot. Looking down, I saw a rattlesnake crawling across the top of my foot. I did not move and the snake crawled on. I think I was too startled to move.

One year father told me if I took care of the bees for the summer, I could have all the money from the honey that was sold. I was very happy at the end of the summer, because I received about $700 for the honey, which was enough money for a new car.

* * *

Marie had a friend named Dorothy Heterbrink. They were high school Juniors, when Dorothy came to the ranch to spend the summer with Marie. Dorothy told me what it was like to spend a summer at the Tinnemaha Ranch: "I will never forget the stream of water running through the yard and my good friend Marie. We were both juniors in high school at Fullerton. I spent the whole summer at the ranch in 1925. I remember my folks coming up late that summer and saying I was overstaying my visit, which I was, but I was having such a good time.

"Marie and I were two young girls. Mr. and Mrs. Alcorn took us to all of the dances. There were many men who wanted to get acquainted with us, because there were few girls in town. I will never forget the time when a white haired boy came up to the house on his white horse. We didn't know what he was doing up at the house, but later on, we found out. He heard that there were girls at the Tinnemaha Ranch, so he took it upon himself to come out and pay us a call. He came to Fullerton to see me again after that. Marie and I talked about him many times and wondered what had become of him.

"Marie and I didn't pay any attention to the men at the bunkhouse, except for one, a sheepherder. While investigating the ranch, we saw this little shack. The shack was full of dirt and the sink was full of dirty dishes. There was so much dirt that Marie and I thought we would be good samaritans. We cleaned the shack spick and span. It wasn't long after that, we each got a box of candy. Oh! We thought it was wonderful, but to our disappointment, it wasn't from the sheepherder, it was from Marie's father. He told us that he wanted to get each of us a box of candy for cleaning up the messy sheepherders camp. We sure wished that the candy had been from the sheepherder!

"I had never ridden a horse, until I stayed with Marie. We explored the ranch on horseback. I came home sore many times, due to the fact that I was not used to riding.

"I remember picking strawberries every morning and helping with the dishes at the ranch. Marie and I went to the dances that were held at Aberdeen or a place called Keough Hot Springs which had a swimming pool. Marie's parents were very nice to us. On Saturday evenings, they took us to where there was a little action. We both tried to learn to tap dance from the big Victrola in the living room. We kept that phonograph going almost all the time, usually with the tune of 'Little Eyes of Jade' playing. That was a wonderful summer for me which I will never forget!''

The Tinnemaha Ranch was located below the Sierra Nevada Mountains. The fruit trees and the green fields with the high mountain peaks in the background were spectacular. Some of the mountains that surrounded us were over 14,000 feet high. I have often wondered what effect the mountains have on a person. The last time I was in Owens Valley, this feeling was the same as ever. ''Just look at that view,'' I thought as I looked up at the mountains, ''With a view such as that, it could not help but have an influence on a person. I can not imagine that anyone who lives under these noble peaks with such beautiful scenery could help but be happy. This valley would probably have been even more beautiful if they had not taken the water away.''

Chapter Four
Fish Springs School

This chapter contains stories about Fish Springs School. The information is from the Alcorn children and Miss Gladys Shumaker Levering who taught at Fish Springs School.

Fish Springs School was a country school located about seven miles from Big Pine and two miles from the Tinnemaha Ranch. While living at the Tinnemaha Ranch, all five of the Alcorn children attended this school.

Fish Springs School was a one room building with two outhouses, one for boys and one for girls, which were located behind the school. The school house was heated by a coal/wood stove. A stream ran along the north side of the school yard for our water supply.

The first teacher at Fish Springs School that Velma, Lloyd, and Marie had was Mrs. Mary Connors, who also had children that attended this school. She lived about a mile from the ranch. Velma, Lloyd, and Marie were at Fish Springs School in the spring of 1916 when Velma graduated from the eighth grade. This was Mrs. Mary Connors last year at Fish Springs School.

The school board was looking for a school teacher when Miss Gladys Shumaker from Los Angeles heard through her friend, Sarah, that a school teacher was needed at Fish Springs. Sarah married Vincent Burdick from the Red Mountain Fruit Ranch, which was later called the Tinnemaha Ranch. Miss Gladys Shumaker applied and was accepted for the teaching position at the Fish Springs School for the school year of 1916–1917.

The following material was taken from a letter written by Mrs. E. Gladys Shumaker Levering to Lloyd Alcorn about her travels from Los Angeles to Big Pine, then on to the Tinnemaha Ranch where she lived while she was teaching at Fish Springs School.

Miss Shumaker wrote, "The way to travel the 200 miles was by train. About one week before the opening of school I bought a railroad ticket. The understanding was to get on a broad gauge train in the evening at eight o'clock, no sleeping cars were on that line. The train arrived at Mojave at 2:00 A.M., only a quiet, peace loving night station agent listening to the rat-a-tat-tat of the telegraph equipment was heard. After a one hour layover, I changed to the train which ran to Owenyo at the foot of Mt. Whitney. Daylight crept in early. After the sun came over the horizon, it was so warm that it seemed like it was noon. The big steam engine with the melodious whistle rolled into Owenyo. Mr. and Mrs. Geo Brown were in charge of the eating place at Owenyo, Mrs. Brown (Della) was my aunt.

"It was here at Owenyo that I saw for the very first time a narrow gauge railroad. That was not all, because on this narrow gauge railroad sat the most interesting little steam engine, with a coal car followed by two or three passenger railroad cars. Each end of the passenger cars were open. Of all things, the narrow wooden slats made very comfortable seats. The large windows were clean, clear glass. At one end was a coal or wood burning pot bellied stove. No fire was needed since is was a warm August day. An engineer, a fireman, a brakeman and three or four passengers were present.

"In order to make the trip up toward the north worthwhile, three or four box cars were attached to the back. Toward the west stood majestic Mt. Whitney, 14,495 ft. high. Toward the east, the White Mountains Range looked low, smooth, and round. There was a small town called Lone Pine about twelve miles from Mt. Whitney. Owenyo is the train station for Lone Pine. The next stop for this little train was Zurich which would be the stop for Big Pine. There were no fences around the farms nor along the track, but at the same time there were many range cattle grazing here and there. When the passengers least expected it, the train would come to a sudden stop. Often after having filled their stomachs, the cows would lie down between the rails to chew their cud. No animal wants to be run over nor does it want to move fast; the crew put on the brakes, so the animal could meander off if so inclined. If the animal had no fear or ambition to leave the resting place, the fireman must climb down from the engine with a stick or switch he had picked up and chased the lazy bones away. The train pulled and puffed to the station to a smooth stop, after about a half day's travel we were in Zurich. There I met my friends with whom I intended to stay with.

"The school I was to teach at consisted of one room in which eight grades were taught. The year I taught at Fish Spring School there were 25 students, 9 were Piute Indians. The Indian agents made sure that the Indians children did not miss any school. The children were taught spelling, reading, writing, arithmetic, language and geography. Art seemed to be natural to everyone, especially for the Indian children. At that time Mrs. Clarke was superintendent of the Inyo County schools.

Marie and Ray on there way to school riding "Buck." Note the tin lunch pail also Ray with his pockets full of apples and one in his hand.

"The first day of school was fun. I had made arrangements to ride with Lloyd Alcorn and his sister, Marie, who was eight years old. I met them at their gate each morning at 8:00 A.M., for our trip to school. A horse, named Buck, was hitched up to the buckboard. When we arrived at the school the horse was unhitched for the day. Lloyd was always kind to Buck. He then hitched Buck up at the end of school and we all went home."

Before I, the author, was of school age, sometimes I would visit the school with my sister. Marie rode Buck and I would ride behind her. We took our lunches in a tin lunch box. One time, I took my doll when I went to visit. We were riding along and I dropped my doll, the horse jumped and I fell off into some sagebrush. A piece of the brush went through my upper lip and into my mouth. It made quite a wound and I still have the scar to this day.

Miss Shumaker wrote, "The road going to school from the ranch was about two miles. One morning going around Poverty Hill we met an automobile. It was a tight squeeze. Our wheels on the buckboard wagon had to go up the bank and the automobile had to stop until we got by, because he was to close to the edge of the gully. We were on such a slant, that all of my papers slipped off of my lap. If it was cold when we arrived at the school, Lloyd would build a fire for us in the wood stove. One of the boys swept out the school room each evening.''

During Christmas the mothers and Miss Shumaker knitted red socks for the students, filled them with candy and nuts, and tied the top with a ribbon. Mr. Rossie came over with his accordion to play for us. We sang Christmas songs and someone read the poem, ''The night before Christmas and all through the house not a creature was stirring not even a mouse. . . .'' Christmas came and went with lots of fun.

Some of the social events to raise money were dances, school parties, or sometimes the ladies made lunches. The lunches were put into decorated shoe boxes. These boxes were then auctioned off to the highest gentlemen bidder. The ladies' name was in the box and the highest bidder got to share the lunch with the lady. The bidding on the popular ladies lunches was quite brisk and fetched a few dollars. Marie commented, ''I invariably got to eat with the oldest man there.''

Miss Shumaker wrote, ''During the week of teachers institute, there was a big snow. Many new men were brought into town on account of the terrible snow storm we had during the night. All of the high standards (poles) that held the high tension wires had been bent double or blown over. I found someone to take me as far as the school. When I arrived at the cold, forsaken, looking school house, some tramp had gotten in and made himself a nice warm fire by burning some of our books, some old and some new.

''By Valentine's Day snow was still on the ground. The school being seven miles away from Big Pine, we could not get to town to buy Valentines. We made Valentine's for each other with our own poems or notes. They always seemed to mean so much to everyone. In the evening there was a dance at the school. Mr. Rossie played his accordion.''

Aberdeen School was often used as a social center, because it was larger than Fish Springs School and more people lived in that area. This school was about six miles south of the ranch. I remember listening to June Shelley play the saxophone at Aberdeen School. She was a young lady that was often called upon to entertain people with her saxophone. I went to this school in about 1922 with my folks to listen to the first radio. It was an Atwater-Kent radio. Much static and crackling was heard, but one could still understand the man talking and the music playing. Everyone commented it was unbelievable that they could send noises through the air without telephone wires. One person said, ''Oh my, what they won't think of next.''

Early in March, at Fish Springs School on Arbor Day, we planted trees. The children brought long twigs from different kinds of trees. Many of them

A picnic on the last day of school in a grove nearby, with all of the children and parents present.

grew. Lloyd said, "I remember digging a lot of those holes and planting the trees. After 70 years they are still growing to this day."

Before we knew it, it was spring. Lloyd's exams came early since he was the only eighth grader, he brought his lantern with him that morning to study before school started.

On the last day of school Miss Shumaker had decided to have a picnic in the grove close to the school house. All the children's parents came over with their picnic baskets. They had brought sandwiches, cookies, and cakes. We ate all we could, then we picked up and started for home. We had a good time at our picnic.

* * *

When I started first grade at Fish Springs School in 1918. The teacher was Miss Alice Wilson. I rode behind Marie on Buck. I always thought of myself as having been a well behaved student, until I saw one of my old third grade report cards. Under the section of deportment I had a C. The grades went

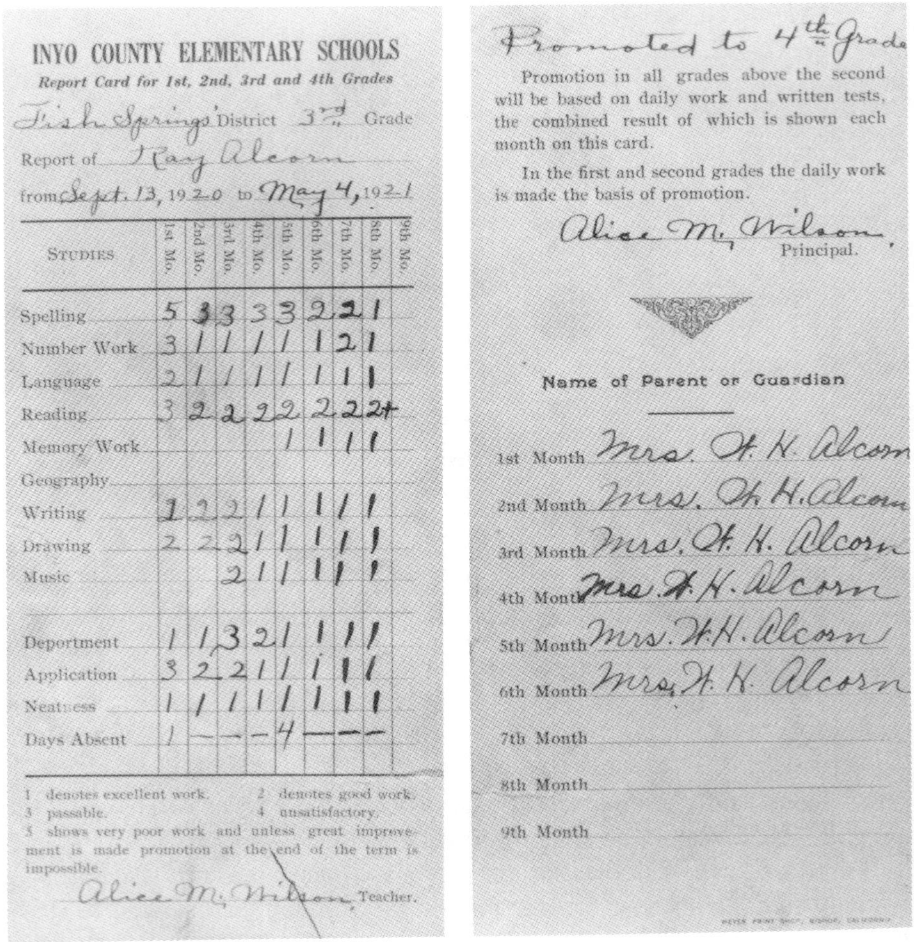
INYO COUNTY ELEMENTARY SCHOOLS

Report Card for 1st, 2nd, 3rd and 4th Grades

Fish Springs District 3rd Grade

Report of Ray Alcorn

from Sept. 13, 1920 to May 4, 1921

Studies	1st Mo.	2nd Mo.	3rd Mo.	4th Mo.	5th Mo.	6th Mo.	7th Mo.	8th Mo.	9th Mo.
Spelling	5	3	3	3	3	2	2	1	
Number Work	3	1	1	1	1	1	2	1	
Language	2	1	1	1	1	1	1	1	
Reading	3	2	2	2	2	2	2	2+	
Memory Work					1	1	1	1	
Geography									
Writing	1	2	2	1	1	1	1	1	
Drawing	2	2	2	1	1	1	1	1	
Music			2	1	1	1	1	1	
Deportment	1	1	3	2	1	1	1	1	
Application	3	2	2	1	1	1	1	1	
Neatness	1	1	1	1	1	1	1	1	
Days Absent	1	—	—	—	4	—	—	—	

1 denotes excellent work. 2 denotes good work.
3 passable. 4 unsatisfactory.
5 shows very poor work and unless great improvement is made promotion at the end of the term is impossible.

Alice M. Wilson, Teacher.

Promoted to 4th Grade

Promotion in all grades above the second will be based on daily work and written tests, the combined result of which is shown each month on this card.

In the first and second grades the daily work is made the basis of promotion.

Alice M. Wilson, Principal.

Name of Parent or Guardian

1st Month Mrs. H. H. Alcorn
2nd Month Mrs. H. H. Alcorn
3rd Month Mrs. H. H. Alcorn
4th Month Mrs. H. H. Alcorn
5th Month Mrs. W. H. Alcorn
6th Month Mrs. H. H. Alcorn
7th Month
8th Month
9th Month

Ray's third grade report card.

by months. The rest of my deportment grades were A's except for a B and that C.

I learned many lessons while going to school, not to mention a lesson in psychology. A friend and I were chasing a pig that had wandered into the school yard during recess. We chased that pig well through our recess and suddenly we looked up and there was Miss Wilson, ringing the bell on the school house steps. Miss Wilson always wore khaki colored mid calf skirts and she had glasses. Her brown hair was always pulled back in a knot behind her head and she had listerine breath. She was only a medium-sized woman, but to a little kid she looked mighty big. She even looked bigger to a kid in trouble. I do not remember her being mean, but at that moment she looked mean.

We headed straight back to class where Miss Wilson was waiting to give us a spanking. She turned my friend over her knee first and paddled him. It did not look like he was being paddled very hard, but he sure was crying and carrying on. The paddle was a board which was about three inches wide and twelve inches long. I decided right then and there, that there was no way I would act like that. Bawling in front of the class was not the way a man should act and I was a man. My turn came soon enough and Miss Wilson turned me over her knee and started the paddling. It really didn't hurt too bad. I did not make a sound the whole time. When she finished, I stood up and told her right in front of the class, "That didn't hurt." That was a big mistake! She promptly turned me over her knee again and proved that she could hit hard enough to make it hurt. I understood then, why my friend had cried and yelled so hard. There were times when being macho just wasn't smart. Ever since then, I hollered like crazy whenever it looked like I might get hurt—even a little bit.

On one occasion when Tiffany Hill and I were walking to school, we saw

Ray and Earl ready for school.

something move as we approached a pile of boulders at the bottom of the hill. To our surprise, it was a skunk. We ran after it in hot pursuit as it ran into a pile of rocks. We spent about five minutes trying to dig it out, but the smell got so strong, we gave up and went to school. On our arrival, the teacher immediately sent us home and called after us, ''Don't come back until you get that skunk smell off of you.''

Us boys went barefoot all summer. Soon our feet became tough. We were able to run through rocks and gravel without feeling any discomfort. We liked to go barefoot, but the rule set down by our folks was that we were not suppose to take our shoes off until the first of May. As I went to school, on such a nice day in the latter part of April, I took my shoes off and stashed them in a rock pile. Lo and behold, that afternoon, it started snowing. By the time school was let out, there was an inch of snow on the ground. The folks often came to pick us up if the weather changed. Since I did not want to get in trouble, I ran as fast as I could to where I stashed my shoes and put them on. Lucky for me they did not meet me at school that day.

Earl, the youngest of us Alcorn kids, also attended first and second grade at Fish Springs School with Miss Alice Wilson as his teacher. After Earl

Marie, Earl and Ray going to school; after Earl started school all three of us rode Buck.

started school, sometimes all three of us kids rode the same horse. Buck was one of our horses that went to school every day. He was a tame horse until it came to cinching up the saddle. He would then try to bite the person saddling him. One time, he reached around and bit me on the cheek. I told my folks about it. Mother was always the doctor of the family, she looked my face over and said, ''Well, I guess your going to be alright because it didn't break the skin.''

Dad asked, ''Why did Buck bite you?''

I told him, ''I was cinching up the saddle, and he reached around and bit me.''

Dad replied, ''I will break that horse of that. Come with me. We will saddle that horse up right now.'' We put the saddle on Buck, and as I went to tighten the cinch on him, he turned around to bite me again. Dad was standing there with a switch made out of a green willow. When Buck turned to bite me, dad hollered and hit him. Buck never tried to bite me again.

Another trick of Buck's would be to swell up, so the cinch never tightened. After you got on, he let the air out of his lungs and the saddle became loose enough to turn and dump the rider on the ground; and some people say that horses are dumb. One weekend, I was riding Buck when the saddle turned and I fell off. I hit my head on a rock. When I hit the ground ol' Buck stood there looking at me. It knocked me half out; the folks kept me in bed for three days to recover. I still have the scar on my forehead from that fall.

* * *

Once a man gave a talk to our class on what cigarette smoking does to your lungs. He had a filter on one end of the cigarette and after the smoke went through it, he showed us the filter. He pulled it apart and said, ''Now, look at all of that gooey stuff, it is caused from cigarette smoke. This smoke goes down in your lungs. Do you want to put that in your lungs? If you smoke, it cuts down on your breathing, so you will have difficulty walking in the mountains and hunting. You may not even be able to walk up the creek and fish. It ruins your lungs.'' Boy, that sure impressed me; I never smoked a cigarette in my life.

Another time at school, the boys told me to meet them in the outhouse. When I got there, they had some wine that one of them had taken out of their dad's cellar. They tried to get me to drink the wine. I would not do it. My folks told me about such situations, they said I should never feel pressured into doing anything that I felt was wrong. It was the same as smoking.

Velma, Earl and Marie posing for the camera.

Marie took the seventh and eighth grades in one year, because she was the only one in the seventh grade. She graduated on May 26, 1922. They closed the Fish Springs School in 1923 and bussed us kids to Big Pine. I'll bet Buck was happy not to make that trip every day.

The following excerpts are from a letter written by Gladys Shumaker Levering to Lloyd:

"December 26, 1978

Dear Lloyd,

The rich experiences I had in Owens Valley have lived with me all of these years.

The school was in the most beautiful part of Owens Valley, at the base of the most picturesque mountains in the Sierra Nevada Range, or so I thought.

It was most gracious of you and your parents to provide transportation for me throughout those many months. After all, you were only about 14 years old and I was 11 years older.

I remember Ray as a little fellow about three years old. Marie and I were pals. And you were one of us because you drove Buck and made the fires each morning. You had to lift Marie up onto the horse's back each morning after

I was on. We could not have gotten along without you. Your other sister, I did not know.

On special occasions we would invite Mr. Rossie to bring his accordion and play for dances. People loved to come. The only light we had was from the lanterns that a few people brought and hung on the walls. The seats were on cleats or straps of board so they could be pushed back to the walls. Our first party was Halloween and the next one was just before Thanksgiving.

Do you remember Fred Lockhurst who had a beautiful voice and played a small accordion and a harmonica. He was a British Canadian.

When the weather became good and everything dried off, we had a few Sunday School meetings at the school. There were no churches or ministers around. Since I played the organ, a little, we sang some hymns and read some psalms. One of the men with a beautiful voice sang solos for us. After the Lord's Prayer we had a little social time.

Although over sixty beautiful years have come and gone the whole experience of living in the great Owens Valley seems as though it were only yesterday.

I could go on and on reminiscing, but I must close for this time.

With love,
The Teacher
Gladys Shumaker Levering''

Chapter Five
The Boys

This chapter is about Bill Banta's family moving to Owens Valley and his work on the Tinnemaha Ranch. Bill Banta and Lloyd Alcorn discuss their experiences together as kids and the people associated with the valley.

It was August 12, 1982, when Bill Banta, Lloyd Alcorn, and Ray Alcorn met in Lee Vining, California to reminisce old times.

Ray: "Bill, when did you first come to Owens Valley?"

Bill: "In 1906, when I was five years old, I came with my parents and Uncle David Ripley and his family. Our family drove a small wagon, drawn by horses. Uncle Ripley had a team of mules—they were slow—they only went about twenty miles per day. A team of horses can travel 25 to 30 miles per day. A person did not get very far traveling back then, compared to nowadays.

"I do not remember much about the trip because I was so young, but I do recall a horse being tied next to a creek with high water. Somehow the horse got tangled up in the rope and fell into the creek. I was scared that the horse might drown, but Uncle David and dad got him up and out of the creek.

"We came to this country through Mono County. We rode by Mono Lake where the highway is now. When we arrived in Owens Valley, we settled in Lone Pine for about a year. Later dad bought a piece of property in Sunland not far from your folks. We raised vegetables to sell. All of the gardening was done with a horse drawn cultivator. All of us kids had to work hoeing weeds. We took a team and wagon to peddle the vegetables in Bishop. Dad went to several stores and from house to house selling vegetables. Sometimes he parked the wagon behind Fred Hooper's store to sell his watermelons and tomatoes.

"One time father got a load of wheat in trade for vegetables. He took the wheat to the bar mill, which was run by a old timer by the name of Kelso. He

Lloyd Alcorn and Bill Banta standing next to one of the pillars that were at the entrance to the Tinnemaha Ranch in 1984.

ground the wheat into flour and took a percentage of the flour for his mill work. It seems like we got a lot of flour back including the bran. Our family ate a little bran, but mostly we fed it to the pigs in the way of slop. Now, by golly, we think it's health food.''

Lloyd: ''Bill, didn't you live in Long Beach, California for awhile each year?''

Bill: ''Yes. Dad thought it was too cold in Bishop, so we started going south in the winter. Dad took us to East Highland or Long Beach because the climate was warmer. It took two weeks to get to Long Beach in the fall and two weeks to come back to Sunland in the spring, so we missed four weeks of school every year.

"I saved some money one summer and bought a bicycle from a young fellow, by the name of Lawrence Bradey. I rode this bike to Southern California. I rode the bike until I was tired, then I would put my bike in the wagon while I sat and rested. I walked my bike through the sandy spots. Most of the road was in good shape, so I rode my bike without a bit of trouble."

Bill: "By the way, Lloyd, do you remember Cecil Thorington?"

Lloyd: "Yes, we were in Ag Mechanics class together. We would supposedly be working on our projects, but when the teacher left for about ten to fifteen minutes we would start to wrestle. Cecil was a lot bigger than I by about ten pounds, but I still gave him a bad time. We wrestled around and things would get stirred up, but you know the teacher never did catch us. We always quit at the right time. Cecil later became sheriff of Mono County."

Bill: "I remember your dad as well as mine knew how to kill a skunk, so they would not smell."

Lloyd: "When I shot them they always stunk up the area. Skunks are funny things, I caught quite a few every summer. I set the traps in the corn field. The reason we had to kill them was because they destroyed our garden and corn fields. They pulled the ears right off of the corn stalk."

Bill: "One story that I want to tell is about Lloyd and the skunks. Lloyd trapped skunks, you know, and after a while he became permeated with that skunk smell. He wouldn't notice it, but everybody else did. In school one morning the teacher said, 'My goodness whose been messing around with skunks? Is there a skunk in this room?' She sat there fussing and old Chris McMillan said, 'Well, don't be looking down at us guys, it's none of us.' Mac, Fuzzy Ray, and the other boys knew it was Lloyd. Lloyd kept a straight face during this conversation, he didn't even dare smile."

"I remember another time when you got yourself into trouble. We always were buddies and still are pals. Anyway, Lloyd came to school one day and said, 'HaHaHa, yesterday was my birthday and you guys never got to spank me.' Well, all the kids, jumped on him and Lloyd looked up at me for help. I said, 'No, you brought this on yourself.' I didn't think I should butt in. Lloyd fought with the kids and held them off for a long time, but they finally got the best of him. One grabbed him by one leg and another grabbed the other, when they got him all stretched out they gave him the worst spanking. Boy, Lloyd, could hardly stand let alone walk back into the school after they were finished with him. He sure fought hard. I wanted to help him too, but it wasn't the right thing to do. I didn't spank him, I let the other kids do that. It was a wonder that some of those sissy guys didn't get into big trouble after

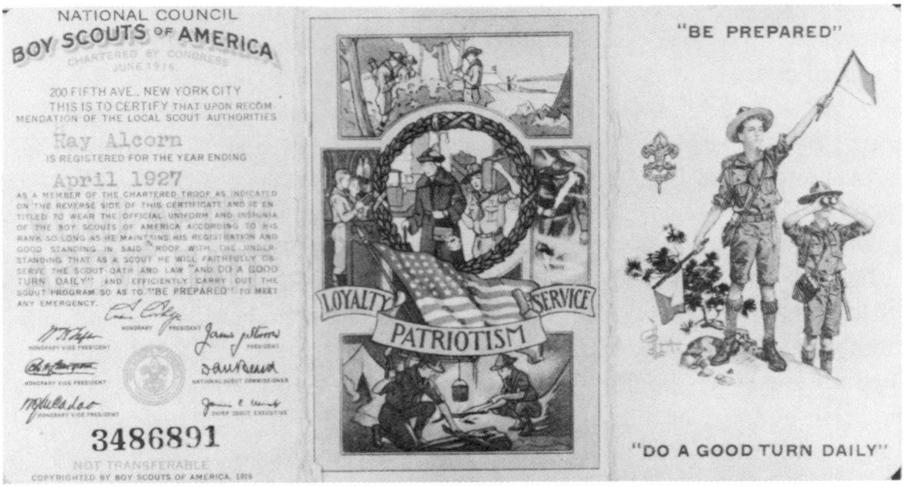

Ray Alcorn's Boy Scout Card for 1927.

that from Lloyd. I guess since Lloyd brought it on himself, he left them sissy kids alone.''

Lloyd: ''Bill, do you remember our Boy Scout Troop?''

Bill: ''Yes, we probably had the first Boy Scout troop in Northern California. As I understand, the Boy Scouts of America started in 1912. About 1912 or 1913, our Northern California Boy Scouts of America started.''

Lloyd: ''The President sent us a little card, with his signature on it. I had that card for a long time, but I don't know if I still have it.''

Ray: ''I still have my membership card from 1927.''

Bill: ''There was a fellow from the Philippine Islands, who had been in the service. He started our Boy Scout troop, his name was J.G. Bryan, Scout Master. Since, Lloyd and I were the two biggest kids, we were named patrol leaders. We went out one afternoon in the hills west of Sunland. When we started back it was getting close to dark and a lot of the little kids were acting tired. They were moving slow saying they could not go any further. Lloyd started telling about the mountain lions and how mean they were after dark. He told them, that in the daytime the mountain lions wouldn't bother you, but the darker it got, the meaner and braver they got. Pretty soon those kids were running over our tracks and stepping on our heels.

''Another time we all went fishing with the patrol master. Mr. Bryan, who had two children, Clyde and Bessie. Clyde was in the same grade as us, but not quite as old as Lloyd and I. We went up Bishop Creek, our camp was above Intake Two. The three of us, Lloyd, Clyde, and I were the scout guides, we were going to sleep next to each other. It was cold that night so Clyde told us, 'I'm going to bed with my shoes on.'

We told him, 'You better not, if you do, we will push you out.' We got ready for bed. I do not remember if I was in the middle or if Lloyd was. We had it all planned out. The one in the middle would get on his hands and knees and the one on the edge would get the covers between his legs and hold them tight. The only thing left was to push and roll Clyde right out on the wet grass. Pretty soon Lloyd gave the O.K. and we gave Clyde a big push, out he rolled, right on the wet grass. Our plan worked better than we realized. He was on a downhill slope, so with his covers being loose he really rolled out of there. He started hollering and crying. Mr. Bryan wanted to know what was going on. I told him how Clyde had gone to bed with his shoes on and how we had warned him if he did, we were going to push him out. Mr. Bryan's reply came as a surprise, he said, 'Well good, good for you, that's just what you have to do, make him stay out a little while.' Oh shoot, sometimes I wonder why we remember things like that.

"I especially remember one trip, when my dad, Abraham, Mr. Alcorn and us boys went fishing at Bishop Creek. Lloyd and I rode the horse, old Barney, behind our dads who were in a buggy. We would get behind and old Barney would run to catch up. When Barney ran, it made Lloyd laugh, the same laugh he has now. Well, by golly, Mr. Alcorn and Lloyd had not fished in Bishop Creek very much. But a person would have thought that Mr. Alcorn must have fished there a lot, because he caught approximately 40 to 50 fish the first day. They were little trout up to a foot long. It was at Dutch John's Meadow below the power plant. After we got the fish home, we salted them. We were allowed twenty pounds or fifty fish a day. On many fishing trips we caught our limit.

"Oh, going back to Cecil Thorington I remember he used to fish with minnows at Owens River, in Long Valley up around Hot Creek, and catch big fish. He was an expert at fishing with minnows. Cecil was a great fisherman. I would buzz around with Cecil and the other kids my age. We were always hunting and camping together."

Lloyd: "Cecil Thorington went fishing someplace. I don't know where, but he brought back two big fish. I still do not know where he caught them, he never would tell me. They were two big trout, they must have weighed, oh, I'd say five pounds each."

Bill: "I remember when Lloyd took his car with Fuzzy Ray, Cliff McMillian, Frank Miller, and myself and we all went fishing. It was around the first of May, the very first year they had the Rainbow Club. Lloyd and I hiked up to the big hole at the gorge. We caught our limit of trout and entered them in the Rainbow Club contest. Lloyd took first prize and I took second. First prize was a rod or something and second place was a button. I still have that Rainbow Button.''

Lloyd: ''Well, our winning first and second prize entitled us to an all time membership to the Rainbow Club. Boy, those trout were beauties, they were nice trout.''

Bill: ''That was the same night we were arguing whether or not we could fish in the moonlight. A bad report got out about us. Fuzzy Ray had made a bunch of hooch, in a five gallon can. It was supposed to be wine, but it did not turn out, because it was not mature yet, so there was no kick whatsoever. Us kids were singing, playing, and hollering until midnight. It got reported that a bunch of school kids were drunk up there.''

Lloyd: ''Yea, we had a little tube that we put into the keg and we would drink the hooch by sucking it out. The next morning, Fuzzy Ray said, 'We need to have another drag on the hooch.' He got a big mouthful, spit it out and he said, 'It doesn't taste right.' When we looked inside, there were about a million little ants in the hooch.''

Bill: ''They sure accused us of being drunk. We were not a speck drunk. We were just up until midnight having fun. We still got up at daylight and were the first ones up to the creek. There wasn't anybody around the big hole when Lloyd and I got there. Oh heck, I think each fish we caught was about a pound and a half, nice big ones. Twenty-five was the limit then.''

Lloyd: ''I remember when Cliff came to the Sunland ranch and we decided to make some brew. We hid it in the wood pile so it would mature. Before it matured, father found it. With a firm voice, he said, 'You get rid of it. And I think you should find better company to keep.'''

* * *

Ray: ''Bill, what did you do when you went to work at the Tinnemaha Ranch?''

Bill: ''Mr. Alcorn offered me a job pruning fruit trees. I was 18 or 19 years old. I told him I learned how to prune trees in Colorado. He started me with a few trees around the south side of the house, which was about 4 or 5 acres. He took me out and said, 'start pruning.' After I pruned the first tree, he turned around and walked off. Mr. Alcorn did not say good, bad, or anything.''

Ray: ''He must have liked the way you pruned the tree.''

Bill: ''I guess so. The man that taught me how to prune trees in Colorado was an orchard man by trade. He must of been from the same school as Mr. Alcorn. The young trees did not need a lot of pruning. I wanted to earn my wages for the boss, so I worked hard pruning those trees. I would walk up to a tree and decide which limbs I was going to cut off first.

"One day, Mr. Alcorn said to me as I was on may way to prune the hill orchard, 'There is one delicious apple tree, when you find it, bring a little sprig home.' A couple of weeks later, I walked up to a tree and said, 'Ah, there is the delicious apple tree I have been looking for.' About five or ten minutes before quitting time I cut a twig off and took it home to him. I told Mr. Alcorn, 'Well gee, it took me a long time, but I finally found it.'

"I recall when Mrs. Alcorn made our lunches, Marie, Earl, Lloyd and Ray were all there. When Mrs. Alcorn said, 'Billy, do you have enough to eat? I keep putting more sandwiches in your lunch and you clean everything up.'

'Well,' I said, 'your dog, Shep, knows about the time I eat my lunch; he shows up everyday a little before lunch. I always divide my lunch with Shep.'

She told me, 'Shep gets enough to eat, so you do not need to feed him.' After that, she cut down on the number of sandwiches I received in my lunch. I still divided with him a little bit. Mrs. Alcorn gave me so much food, I could not eat it all."

* * *

Bill: "I never will forget the day I became snow blind because Sy Bevins and I helped your dad catch seventy-five cottontail rabbits one afternoon in the bright sun when there was about eighteen inches of snow on the ground. Because of my being snow blind, the next day your dad said that I could help with the lambing. That was the most miserable day I ever had. I got hay fever from the alfalfa and dust flying around. I guess that was the best thing for my eyes on account of bringing the tears to them and washing them out. I could not stand any light at all. Even the lanterns and flashlight that we used bothered my eyes.

"During lambing, if one ewe had twins and another ewe lost her lamb, we would give the mother who lost her lamb one of the twins. If she did not want it, we would have to skin her dead lamb and put the skin on one of the twins, so she would think it was hers when she smelled it. Usually by the next day, we could take the hide off and the mother would claim the lamb. If the ewe does not take the lamb, the lamb is called a bummer and has to be fed from a bottle.

"I remember when I got sick at the ranch and if it had not of been for your mother, I think I would have died. Boy, I sure was sick for a few days. Mrs. Alcorn took care of me like I was her own son. I really appreciated her care.''

Ray: ''Bill, do you remember the quail on the ranch?''

Bill: ''Yes. There were a lot of Valley Quail on the Tinnemaha Ranch, when I worked there. According to some sporting goods magazine, quail were too rich to eat everyday. So five of us thought we would try it for a month. There was George Bryan, Sy Bevins, a guy by the name of Shorty, Mr. Alcorn, and myself. We killed the quail. Mrs. Alcorn and Mrs. Smith, who was helping cook at the time, cooked the quail for us. We had quail fried for breakfast; we even had Mrs. Smith roast them and cook them various other ways. Before the time was up, George and Mr. Alcorn got the flu and they ended up not eating much of anything.

''We were suppose to pick Shorty up in town one night and by golly, we could not find him. When he finally got home, we informed him his quail was in the warming oven if he wanted to eat it. He said, ''The heck with it, I'm not interested.'' He never made it, but he could have. So, out of the five of us, Sy Bevins and I were the only two that ate quail every day for a month. Quail as well as cottontails and jackrabbits were thick on the Tinnemaha Ranch.

''I remember one particular incident with Sy when he was hunting. Some people, when hunting and see a deer, become very excited. They shake so much that they can not hold the gun steady enough to sight the gun properly. We called this condition, ''Buck Fever.'' In extreme cases of ''Buck Fever'' the hunter may shoot into the ground before he has the gun to his shoulder. In other instances the hunter puts his gun to his shoulder and ejects the shells without firing a shot. If two hunters are present, the one with this fever will usually eject a shell each time he hears his partner fire a shot. Sy Bevins did this with a deer, he emptied his gun on the ground except for the last shot, when out of desperation, he finally hit the buck.

''I went hunting with a guy in Big Pine, who reminded me of Sy. By golly, I told that guy I would meet him at the big tree. He was from Los Angeles and I knew he could not hike very far. It took him awhile to get up there. I could see the big tree from where I was, when I jumped four big bucks. They slipped away from me and I didn't get a shot. Meanwhile, I slipped around trying to intercept them. They went up to a bluff and by the time I got around them, the bucks had gotten away. When I came back to their tracks, I saw that the bucks had separated. I heard a shot about an hour after I jumped those bucks. I went to the big tree where the guy was standing. The Los Angeles dude told me that a big old buck walked right in front of him and stopped. He told me his gun was shaking bad and he shot at the buck six times. Finally on the last shot, he said to himself, 'I am going to wait till my gun is still, then I am going to pull the trigger. I had been missing him and he was still standing there.'

'I only heard one shot and I barely heard that.' I walked over to where he was standing and just as I had suspected, it was the same thing that Sy had done. All but one shell laid on the ground unfired. On the last shell, this dude shot the buck right through the middle. The old buck was so big, we had to cut him in half to carry him off the mountain."

* * *

Bill: "I worked in the Winter and early Spring on the Tinnemaha Ranch. In the Summer I helped my dad raise vegetables and peddle them. I remember when my dad bought our first car in 1919 for $500.00. It was a Ford touring car. But before long, dad made a truck out of the car by putting side boards on where the fenders were, to make boxes to put vegetables in. He also removed the back seat to make more room for the fruit and vegetables. We also made trips to Mammoth in Mono County to peddle vegetables and melons. I enjoyed those trips."

Ray: "Bill, were you in business with your dad?"

Bill: "Yes, I was in business with dad for about a year before he retired. When I took over I bought a little old truck and started peddling vegetables and melons in the mountains. I had a total of five men working. That was when my business really started.

"I bought a new ton and a half truck for $1500.00. The dealer in Bishop said, he would wait until the crop came in for me to pay for the vehicle. I said, 'No,' because I had enough money in the bank to purchase the truck. It was lucky for me, I spent my money on the truck and did not lose it when the bank went broke in 1927.

"The banks closing down did not affect me as much as it did other people. I made $1700.00 that summer and I continued doing well through the Fall. A lot of the farmers lost almost all they had, because they only got a certain percentage of their money back from the banks. It was sad how some of the farmers had to sell their land for such low prices compared to the actual value of the land. If they could have held on a little longer, those farmers would have made a fortune on their property."

Ray: "Lloyd, did father lose any money when the banks went broke?"

Lloyd: "Father did not lose very much, but I did. I had deposited $1700.00

and two days later, the bank went broke. I lost all of it. Later on, over a period of years, I got back about 30% of my money.''

Ray: ''Bill, thanks so much for visiting with Lloyd and I. It was a lot of fun listening to all of your stories.''

Bill: ''It was a great life and I sure enjoyed it while I was at the Tinnemaha Ranch with Mr. and Mrs. Alcorn and the rest of the Alcorn family.''

Chapter Six
Fullerton

This chapter is about the Alcorn children attending school in Fullerton, California. These stories tell about their stay with Mark and Bess Alcorn, along with the memories of their school activities and friends.

Mr. Charles C. Chapman, owner of the Tinnemaha Ranch, lived at the eastern end of Commonwealth Avenue in Fullerton. My father was on his way to report to Mr. Chapman at his residence, when he saw a mailbox with the name J.M. Alcorn printed on it. Father thought, "I wonder if that could be my cousin, whom I have not seen or heard from since we were kids in Kentucky." He went to the house, knocked on the door, and sure enough, it was Mark. When father was told to report to Mr. Chapman's place in Southern California, he never expected to see his cousin, Mark Alcorn and his wife Bess. They were actually our cousins but us kids called them Uncle Mark and Aunt Bess.

In the 1922 *Pleiades,* the Fullerton Union High School Annual, the faculty pictures show "James Alcorn" and off to one side of the name is written, Uncle Mark. He was an agriculture teacher at Fullerton High School. He quit teaching school in 1922 to start a nursery. He grew fruits, vegetables and cut flowers for the Los Angeles market.

* * *

My folks thought that the kids in the Big Pine School were out of control and not cultured enough. The kids would roll toilet paper down the aisles; in those days, that was considered rough. One night Marie and a group of her friends from the Big Pine High School, stole some melons from father's watermelon patch and started throwing them at cars. The kids hit a windshield and broke it. They never did pin the watermelon episode on the kids, even though, dad found a kids hat in his watermelon patch. The folks said to us, "I think we should find a better school."

Uncle Mark and Aunt Bess suggested that us kids come to Fullerton and stay with them while going to school. They had no children of their own, but since Uncle Mark was a school teacher, the folks knew he could provide the supervision we needed.

Marie was the first to move to Fullerton in 1924. She stayed with Uncle Mark and Aunt Bess, for two years, while she attended Fullerton High School her sophomore and junior years.

Uncle Mark and Aunt Bess were living at 1211 East Commonwealth Avenue. On the north side of the avenue, Uncle Mark owned ten acres of orange and persimmon trees. He also grew nursery stock and vegetables between the rows of fruit trees. The basement was full of canned fruits and vegetables, that Aunt Bess canned to use in the winter months.

Marie said, "I worked for my room and board. I helped by cooking, cleaning, and doing various jobs around the house. Uncle Mark raised chickens to sell, and as a part of my 'job' on Saturdays, I helped select the fryers that were to be butchered. We sometimes killed the old roosters and hens to be sold for roasting or boiling.

"After the chickens were killed, we dipped them in hot water and plucked them. The chickens were gutted, cleaned, cut up, and then wrapped in paper. After this process they were ready to be sold. Uncle Mark and I also gathered eggs to be sold. People would come by on Saturdays and Sundays to buy the eggs and chickens.

"Uncle Mark and Aunt Bess always went to the Methodist Church every Sunday. I did not go to church, but I did go to Sunday School with a group of girls from school.

"Uncle Mark and Aunt Bess were very strict with me. Most of the time I respected Uncle Mark's orders, but one time I didn't. I was invited to go to a Masonic Dance with a girlfriend. It was a dance put on by the Fraternal Order of Masons. These dances were for the community to enjoy. My girlfriend's dad was going to take us, but Uncle Mark and Aunt Bess said I could not go, because they did not believe in dancing. I snuck out my good clothes and went to my girlfriend's house. After I had left, Uncle Mark and Aunt Bess checked my closet and found that my good clothes were gone. They came right over to my girlfriends house and got me. Boy, did I go home in a hurry! Going to a dance was out of the question.

"I told Uncle Mark that my boyfriends were scared of him. One time a boy wanted to take me out, so I said, 'Well, you will have to come over and see Uncle Mark.' He must have been scared, because he pulled up, honked the horn, and did not even bother to come in. Boy, Uncle Mark flew out the door and told him that he could take his car and go home, because that was no way to treat a girl.

"Another time, a boy came to call on me, I was hoping Uncle Mark would change his old chicken manured boots, but he didn't. He came right into the house and sat there watching every move we made. I was so embarrassed!

"I became lonesome for the Big Pine School, my family, and the Tinnemaha Ranch, so I moved back home and graduated from Big Pine High School in 1926. Velma and Lloyd both graduated from Bishop High School, Velma in 1920 and Lloyd in 1921. While in high school, Velma became friends with Geraldine Moser and her two brothers, Jim and Glen. Velma

Velma Alcorn Moser and her husband, Glen Moser.

went to college in Fresno, California in 1921, while Glen Moser was in Kansas City going to mechanics school. In 1922 Velma returned to Bishop and worked as a secretary for a utility company. At this time she became reacquainted with Glen, and in 1923, Velma and Glen were married. They moved to Santa Monica in Southern California.''

* * *

In 1927, I, the author, went to stay with Uncle Mark and Aunt Bess to attend high school. At Fullerton, there were about a thousand kids compared to Big Pine where there were less than a hundred. Coming from a small to a large school made quite a difference to a freshman boy. I did not realize how small Big Pine School was in size until, I saw the Fullerton High School.

I did not think Uncle Mark and Aunt Bess were as strict as Marie did. Maybe because I was only a freshman, I was not particularly interested in going anyplace. Uncle Mark had the tendency to kid quite a bit, but he always was very fair with me. Aunt Bess was a dear lady; when I stayed with them, I was treated as if I were their own child. I went to church with them, and enjoyed Sunday School and the friends I met there.

Like Marie, I worked for my room and board. Uncle Mark went to Los Angeles on most Saturdays, to take vegetables or plants to sell. He left me at his nursery and garden plot to work, while he was gone. I became so tired, that I laid down in the shade. The next thing I knew, he was standing near me saying, ''Ray, Ray, what are you doing lying out in the sun?''

I said, ''I was in the shade when I laid down.'' He always liked to kid me about my long nap and about other things I did.

A few miles from his home in Fullerton, Uncle Mark owned some property at LaHabra Heights where he worked on weekends growing vegetables and fruit trees. It was located alongside of a golf course.

Uncle Mark had ''Keep Out'' signs along the golf course side of his property. Some of the golf balls would fly over into Uncle Mark's vegetables and large tomato patch. Ignoring the signs, the people would come into the garden looking for their golf balls and trample his tomatoes. Uncle Mark and I occasionally gathered up what golf balls we could find and sold them. I do not remember how many we picked up or what price they were sold for.

I helped Uncle Mark pick persimmons, which have to be picked green for transportation purposes. When green, they are too puckery to eat. If I found a tree ripened one, I ate it right on the spot; boy was it delicious! Uncle Mark said to me, ''Say Ray, that is the tenth or twelfth persimmon I have seen you eat. You may have some trouble tonight, if you eat too many." I thought, "Oh gosh, I definitely do not want to have an accident." Luckily, he reminded me to quit eating them.

Uncle Mark having been a former school teacher, knew the mannerisms of kids. One time he said, "Ray, you had better study your math."

I said, "OK." But, the more I studied Algebra the more confused I became. Pretty soon I quit and Uncle Mark asked, "Did you get your math done?"

I replied, "I guess so."

He said, "Well, then explain it to me."

I thought, "Oh my gosh." I tried to explain it the best I could. He told me, "You don't know it. Now this is the way it is. . . ." Uncle Mark spent about a half an hour working with me, so I could understand and solve the problems. When he got through, I'm telling you, I understood all of the next day's lessons. Uncle Mark frequently kept track of my progress with the lessons, he made sure I understood them. I think that is the reason, in later years, I was pretty good at math.

Uncle Mark always encouraged me to do various things and to go different places. Because of his encouragement, I remember going with a group of friends to the La Brea Tar Pits, the ocean, and a tour of the battleship, California. Other places I visited were football games when USC was playing in the Rose Bowl, the New Years Rose Parade at Pasadena, baseball games at Wrigley Field, and occasionally the beach to swim.

* * *

While in Fullerton, my freshman year, I recall that all of the older boys said, "You had better go out for football, better try out for football." I thought, "Well, maybe I ought to try out for football." They gave me a uniform and said, "You guys who have never played before, you get on this line. The ones that have played before, get on this other line." Here we were up against the "professionals." At the call of a certain number, we were to charge. When I charged the guy across from me, he pushed my head right down in the dirt. I thought, "Boy, I do not want anymore of this kind of stuff." After one day, I had enough of my head hitting the dirt. As a consequence, I got involved in swimming and tennis.

My freshman year, I took swimming and in my sophomore year, lifesaving. They gave me a certificate and a patch that could be sewn on my swim trunks for having passed the life guard test. I was really proud of the speed

I achieved while swimming. I went out for water polo as well as other water sports.

Three of us boys thought we were good enough swimmers to swim across Balboa Bay. We started swimming and everything was fine. Soon, one of the boys said, "Boy, I'm not going to make it!" We were out in the middle of the bay and it was scary. We decided to swim to a boat, that was anchored nearby, to rest. We hung onto the anchor chain for a long time. The boy acted like he was afraid to leave. It probably would have been smart if one of us strong swimmers went for help. Instead we said, "Well, we had better get going or we will not make it." We started swimming again. The tired swimmer was going slow along towards the last. When he finally got to the beach, he crawled up on the sand and collapsed. He laid there for a long time before we finally got him up. As I think back now, if it hadn't been for that anchored boat, that kid could have drowned.

When I lived in Fullerton, I played the saxophone in the high school band and in the orchestra. I remember the accomplishments of some of my fellow classmates. To this day, I remember how Charles Cramer played the xylophone at the assemblies and other meetings. I played in the orchestra, along with about fifty-two others. One of which was Richard Nixon, who became President of the United States. Richard played second violin, but he was known more for his oratorical abilities. In the Fullerton Union High School annual *The Pleiades* of 1928, Alice Bloodgood wrote, "One of the outstanding accomplishments was that of Richard Nixon who won the oratorical contest in his sophomore year against two seniors. He also won the sophomore forensic and represented us at the county meet in Orange County."

I remember the band played at the football games. For the out-of-town games we would leave before the team and have lunch. One time, when we stopped for lunch some of the guys put a little piece of butter on the end of a table knife and flipped it towards the ceiling, where it stuck. Man, it would just ZZZZZZZing right up there. I do not remember flipping the butter on the ceiling, but I probably blocked it out of my mind real quick if I did. Good grief, it was a terrible thing to do. Hard to tell what could have happened if they had caught us.

* * *

Mother went to Fullerton and rented a house during the school months of September, 1927 to May, 1928. Marie, Earl and I stayed with her and went to school. Earl was in grammar school, Marie was in Junior College, and I

Ray Alcorn when he played the saxophone at the Fullerton Union High School.

was in high school. In the Summer, we returned to the Tinnemaha Ranch to work.

Marie recalled, "When I was in Junior College, I took swimming lessons. I believe the Junior College Campus and the Fullerton High School were on the same property, but they had different buildings. Mother stayed for only one school term. Being away from home and father was not working out. I was staying with the Heterbrinks the last year of college, when a girl came over from Anderson's Oil Drilling business and asked me if I wanted to go to work. I said, 'Sure.' I went over there, applied for the job and got it. I quit Junior College about three months before graduation to work for Mr. Anderson as a secretary."

* * *

Glen Moser's brother, Jim, played on the Southern California Football team as an end. He was one of the stars. Several times we went to the ball games to watch him play. Naturally, since he was on the team, he managed to get tickets for us.

I remember going to several different places. We went to a place they call Venice and if I recall correctly, this was a pier out over a salt water bay This area was an entertainment center where they had all sorts of games. We did not have much money to play the games, but we went down to the pier several times. In one place, they had a ticket office, where a person could pay a quarter or four bits to go in and see a show. Right where you bought the tickets, there was a hole in the floor. They had the floor adjusted, so that it connected to a source of air. My brother-in-law, Glen Moser, knew about this hole. He managed to get Velma, Marie, or one of the other gals to stand over the opening. The air was triggered by the man that sold the tickets. Of course, when the air came on it blew their dresses way up. This was a great big joke to us boys.

It was always a problem for me to ride to Southern California, because all my life I have had trouble with motion sickness. My mother told me that when I was two years old, they seldom took me into town, because I would get carsick.

I remember going to Southern California with Velma and Glen, when I got real sick. I heard Glen telling somebody, "I got it figured out, that boy is going to get sick no matter what you do. Even if you drive slow, he still gets sick. The best way is to get in the car and drive like crazy. Drive as fast as you can, until he starts hollering to stop. Stop, let him out to walk around and get his bearings. Then drive like crazy, until he starts to holler again."

The second trip I made with Glen, he said, "I know what to give that boy. There is a soda fountain in the drug store in Majove. When we get there, I will get him an alka-seltzer." I recall he had me sit on a stool at the counter and said, "I want an alka-seltzer for this boy." He said to me, "Now, drink this, just slug it down." I no more than drank it down, when it came up like it was jet propelled. I sure made a mess of that counter. He got me out of there as quick as he could. Boy, I was sure glad to get out of there!

It seems like Glen always got stuck driving me to Southern California. I was always asking him to stop, so I could walk around or go to the bathroom. One time, Glen pulled into this service station that had a gravel surface. I started to walk toward the bathroom and passed out. I fell right on my face.

A piece of gravel went through my upper lip. The inside of my lip was bleeding, I was sure a mess. That was not the only time I fell and ended up with a bruised face. I sure had an awful time with motion sickness. It turned out as I got older and was able to drive, I could control my motion sickness.

Someone wanted me to take a car to my sister's place in Santa Monica. Driving along the highway near the beach, I saw a little airport. They were giving people airplane rides for five dollars. I thought it would be fun to ride in an airplane. So, I went over and asked the fellow, "Can you give me a ride?"

He said, "Oh, yeah. Get in." The biplane had two open cockpits. He took my five dollars and belted me in. The pilot was in one cockpit, I was in the other. We took off. We had not been off the ground for more then a minute or two and I could tell I was getting sick. The pilot flew out over the ocean where the waves were coming in. He turned the plane up on its side, to give me a better view of the beach. Man, did I get sick. Signalling to him to take the plane down, he landed right away and said, "Hey kid, don't ever get in my plane again, you made an awful mess." I thought, boy, he doesn't have to tell me to never get into an airplane again.

Ray Alcorn and Clinton Spots in front of Clinton's car.

* * *

I met many people during my stay in Southern California. Clinton Spots and I went numerous places together in his fathers car.

Reading the Fullerton High School annual of 1927 and seeing the names of over one hundred friends who signed my copy of The Pleiades, brings back fond memories. Most of my friends, I have never heard from since leaving Fullerton.

I wonder what happened to my friends of over 60 years ago.

"Where, oh, where art thou?"

Chapter Seven
The Forest Ranger—Lawson Brainerd

Lawson Brainerd has been a close friend of the Alcorn family since 1924. He was a district ranger in the U.S. Forest Service. He developed many trails in both the Sierra Nevada and the White Mountain ranges along Owens Valley. He made a valuable contribution to this area. Lawson Brainerd and his wife, Helen, came to the valley in the Spring of 1924 and left in the Fall of 1929.

Father met Lawson Brainerd when he called at the Brainerd's home to see if he could get a permit to build a fence on the U.S. Forest Service land for his bucks (sheep). Through a long chain of events, a life long friendship developed between the Brainerds and the Alcorns.

I have fond memories of Lawson Brainerd, his wife, Helen, and daughter, Naeda. Lawson had considerable influence on my life. Every so often father and I went to Lawson's home in the evenings to study and learn the names of different birds. We learned the scientific names of all of the ducks and studied the habits and distribution of all of the North American birds.

Lawson gave me fly tying lessons. I was taught to use thread to tie feathers on a fish hook to make various artificial insects. This made it more fun to catch fish. Several times while I was tying flies, Lawson, with the point of his knife, would scratch a design of an unusual critter on my tackle box.

The Brainerds would sometimes invite us over for dinner. Once, when my folks had gone to Los Angeles, the Brainerds invited me to eat dinner with them. I was not very sharp that day, because when I got home from school I ate a lot of food. Not long after that, the phone rang and I heard Lawson's voice, "Ray, aren't you coming to eat dinner?"

"Oh yes, I'll be right over," I replied. They knew I was a big eater so they kept telling me to have more and more, my poor stomach haunted me all night.

Lawson often told the story of a picnic with our family. We brought

Some of the drawings that Lawson put on Ray's tackle box with the point of his knife.

homemade ice cream and other food items, the only thing missing was the fish. Naturally we went up the creek to fish, and soon brought back our share of nice trout to fry. When we got through eating there were quite a few fish and a little bit of ice cream left. Lawson later recalled in astonishment, "Ray took those leftover fish, boned them, put all the leftover ice cream on top of those fish, and then he ate it. My what a sight!"

In 1929, I worked part time with Lawson Brainerd for the U.S. Forest Service. Working for Lawson was interesting and educational. I not only learned about birds and animals, but he also told me the names of trees, grasses, and bugs.

Lawson frequently ate unusual things. One time he said, "Ray, did you know in Asia they eat Locust bugs? I suppose they are different than the Locust here, but we ought to try them." We caught several, pulled their heads, wings and legs off, and tasted them; they tasted like cheese. We ate them raw but they might have tasted better if we had roasted or cooked them some other way. As I remember, those locust were quite crunchy. I think that's the main reason why I disliked them. Boy, I sure did not want to eat anymore of those things.

Lawson was a great guy, he used to always say, "If you are going to do anything, do it right and on time. Don't mess around." I always remember Lawson saying, "Look at those people, sitting around wasting their time."

Father, Lawson Brainerd, myself, and one other man were going to plant trout in a lake high in the mountains. These baby trout were about an inch long. We put the fish in five gallon milk cans and tied the cans on each side of the pack horse. We put ice in the cans to keep the water cool and started up the hill. When it was too rough for the horses, we carried the cans the rest of the way to the lake. This meant each of us had to carry one five gallon can on our back. As we were carrying those fish to the lake for the first time, somebody said, "Well, what is the name of this lake?"

Somebody else said, "I don't think it has a name."

Dad said, "Let's call it Brainerd Lake." On the next trip to the lake, father painted a sign and nailed it to a tree forever declaring it Brainerd Lake. Since that time, it was called Brainerd Lake, although presently Brainerd Lake is misspelled on the maps. They spell it Brainard, with an "a," instead of with an "e."

While working for the U. S. Forest Service, I will never forget one job I had. Lawson took me far back into the Sierras where I was to build a trail. We had left the valley very early in the morning with the aid of horses and pack saddles. No surveys had been made and Lawson's instructions to me were: we want to build a trail from this point up toward that point, or to that tree, then over to another point or pass. Lawson was planning to return that same day. We unpacked the pack animals, put up a tent, he bid me good-bye and was on his way down the canyon.

Although, I was only seventeen years old, neither he nor I had any fear of my being left up there a week or ten days alone. The tools for building the trail were all hand tools consisting of a shovel, a pick, an ax, a mattock, and a long iron bar for moving the large boulders. My camp gear was very simple, besides my sleeping bag, I had a frying pan and a small pot for cooking over an open camp fire. The days that followed were lonesome. There were no lakes or streams nearby where I could fish, so I spent my time working. Each morning I was up and working at daylight. When Lawson returned he was amazed at the amount of trail I had built. I was sure glad to see him.

On another trip, Lawson and I went up to Coyote Meadows during a dry year and discovered there was no running water coming from the spring. The cattle had been around the area where the spring usually flowed. There was only enough water seeping out of the ground to fill their hoof prints. Being thirsty, we decided that the water in the hoof prints was better than no water at all, so we drank it. Needless to say, the water did not taste very good.

I remember one specific incident that happened when I went on horseback up to Birch Lake with Lawson Brainerd and Sam McMurry. After we fished for a while, we went to where the horses were tied and ate our lunch. After

Ray Alcorn ready for one of his trips into the Sierras.

lunch Sam McMurry laid down and went to sleep. Lawson was never one to "ha ha" very loud, but he always had that big grin on his face. He said, "Let me show you how to startle Sam McMurry. Since Sam is an old cowboy he knows what run away horses sound like." Lawson went over to where Sam was sleeping and started slapping his hands on his legs, so that it sounded like horses running. At the same time he was doing that, Lawson started hollering, "Run, Sam, Run!" Well, Sam half-awakened to the sound of running horses and Lawson shouting, "Run, Sam, Run," he took off running. He did not even stop to look back, he just took off through the three foot willows. Of course, he only got about twenty feet when he realized what had happened.

I will not repeat what Sam McMurry said to Lawson, but it wasn't very nice. There were even a few words that I had never heard before. The big grin that Lawson always wore, sort of disappeared after Sam's dissertation.

I went with Lawson over Taboose Pass to Bench Lake. Lawson had a pack mule that was not broke enough to take on a trip, but Lawson decided to take him anyway. After riding several hours high in the mountains, this mule decided to buck. He bucked so hard that he lost part of his pack. Supplies were scattered all over the hillside. By the time we gathered up the supplies, I thought Lawson was going to kill him.

Another time going over Taboose Pass, Lawson had a mule that decided to buck and turn around to go back home. Lawson said, ''We will hurry up and go over the hill, so that he cannot see us and wait there. He will follow us because he doesn't want to be left out here alone.'' Sure enough, when we got over that ridge that old mule came a huffing and a puffing along. He was not about to be left alone.

* * *

Smoke was billowing up from the far side of the White Mountains; Lawson decided we should check this fire out to see if it could be contained. Lawson asked Jack, his brother, who was visiting him to go with us. The three of us drove up the west canyon as far as an old mining road went. We parked the car, took a couple of canteens of water, a few fire fighting tools and then we started walking up the canyon. Within an hour we reached the crest of the mountain and saw the fire below us on the east slope of the mountain. Within half an hour we had built a few fire breaks, by clearing pathways to prevent the fire from spreading. It was a hot day and we were sweating up a storm. After several hours work we drank all the water from our canteens. Lawson decided the fire was contained enough to burn itself out. Since we were out of water, we started back to the car.

After a few hours of fire fighting we were not very energetic. We soon found that it had been much easier to walk down to the fire a few hours earlier, then it was walking back up and over the crest of the mountain. Lawson soon realized that we had to slow down and make more frequent stops to rest. Jack couldn't keep up with us. He wasn't accustomed to this kind of strenuous work and we were all very thirsty. Half way up the crest of the hill, Jack gave out. We discussed what to do and it was obvious that we had to get some water. Our mouths were as dry as cotton and it was my first experience and realization where the old saying, ''I was spitting cotton'' came from.

Lawson Brainerd at Coyote Meadows with his dogs and a nice bunch of Sage Grouse.

We agreed that Lawson should stay with his brother and I would go for some water. I started up the hill and soon reached the crest. Then it was no effort to walk down the canyon to the vehicle and our water. I had previously been told stories about Death Valley travelers running out of water and drinking so much at once that they dropped dead. So, I only took a sip out of the canteen, waited a few minutes, then took another sip and started back up the hill with a full canteen of water. My thoughts were, "I had better get back over the hill and take the water to Jack." Lawson also needed a drink, so I moved as fast as I could to the top. From there down to the men was a breeze. The water revived me; going down hill did not take much effort.

The men hadn't moved since I left, they sure were glad to see me. The hour rest had been good for Jack and with a little water we went slowly up over the crest and down to our car. Years later, Jack sent word via Lawson, "Tell Ray, thanks for that drink of water."

* * *

The rest of this chapter is Lawson Brainerd's story of his experiences in Owens Valley. Lawson Brainerd wrote, "You had to be eighteen to get into the Navy, since I was only seventeen, I lied to get into the service. A Forest Ranger had to be 25 to 45 years old to be in the Forest Service before the war. Since I had military experience I was hoping they would let me in.

"I received the letter from the District Ranger in Truckee. He wrote, 'I need a young fellow, who has a good mind of being alone and isn't afraid to work.' I remember my dad chuckling when he came to that paragraph, saying that is one place you are qualified. You never were afraid to work.

"'You will probably be the youngest District Ranger in the U. S. Forest Service and there will be complaints in the Regional Office that you are too young, but we think we can persuade them. Your work there will be mainly grazing and that seems to be your line. Go home and talk it over with Helen.' She said yes!

"It seemed ages of hope and worry as the promotion zig zagged through the departments, roosting on each desk before it was looked at, questioned and approved, Mr. Bigelow lobbying at every turn. Then the joyous day when the O.K. of the transfer came through.

"We rushed back to Loomis, got out Helen's furniture that had been stored in the hay loft of our home barn since we were married. We brought in the horses from winter pasture; dad had given us another horse, Fossil Face, a big boned, big footed bay mare. In making arrangements for a box car to ship all our things and railroad passage, we were stopped short on the brink of the meanest arroyo that we had to cross.

"To go to Big Pine by railroad, it was necessary to pass by Reno and go through Nevada and enter Owens Valley from the north.

"The scourge of the old world, hoof and mouth disease, had broken out in the San Joaquin Valley and foothills.

"Nevada, whose main economy was livestock, very rightly, wanted no animal from California to cross their line. Telegrams and phone calls crowded the lines. I never did know the details. Somehow, I got permission to load the car. With Pa's help, we loaded the furniture on one end of the car and the horses on the other, then a bulwark of 2" x 4" on each side leaving me a small space in the center. Helen and Hogan, our little fox terrier, were forbidden to go, thank heavens. Mama put up a box of food and Dad put in a sack of rolled barley and a bale of hay. I kissed Helen good-bye and the train chugged east.

"It was a new and novel experience viewing the Sierras from the door; then down the east side looking down on Donner, when the snow sheds permitted. Past Truckee and descending into the Truckee River Canyon. Everything sliding along smoothly until we landed in Reno—Spark's Stockyards. That is when the buzzard hit the windmill!!

"Nevadans were already chewing their neighbors' finger nails with the worry that the disease would flow over the Sierras and infect their herds and vows were made that no bug, bird or beast, except man, should cross the State line. When it was realized that a loosely covered Pandora's box of three horses had appeared in their railroad yards, Mephistopheles strolling through the Vatican on a Cooks guided tour would not have caused near the consternation. My car was immediately shoved off on the most remote siding.

"A pad of officials assembled at a safe distance and one brave soul came to the car and demanded my travel orders. He took them and returned to the group. They milled and gesticulated around the orders for what seemed a long time. I guess the papers were in order because two men came back with them. One of them was an enormous, brawny, gun toting lad with a big star on his chest, the Sheriff, I guess. I was admonished, in terms that I didn't want to mess with, that I was not to get out of the car nor let a straw drop out. Then the group adjourned, I guess to 'make medicine.'

"I had told them that my horses were thirsty and about noon a truck pulled up along side of the tracks with some five gallon milk cans with water and a tub. The driver cautiously helped me lift the cans into the car and said he would bring more the next day. He didn't want the cans back.

"That afternoon a group gathered within talking distance of the car and eying me as if I were a rabid coyote with leprosy spawned by the Devil, questioned me in detail. I plead ignorance and just told them that I was under orders by the Forest Service to take care of the horses from the point of departure to their destination in Big Pine as per shipping orders. I think any of them would gladly have donated their best rawhide lariat to hang me if they thought it would solve anything.

"It had been a warm day and by evening the horses were sweating and doing other natural things about every three hours, the atmosphere in the car was starting to get high.

"Early the second morning, the man who had brought the water (I think he was employed by the railroad), bless him, brought me a bag of sandwiches and a jug of hot coffee, also a jug of drinking water then, about noon, more water for the horses and a bale of hay, also more food for me.

"Cass, my horse, was protesting vocally before the sun had peeped over the desert to start another day. I gave the horses a little breakfast and water, then waited patiently in hopes that the railroad man would show up soon with food, coffee and a little news. He didn't fail me. He had become real important as he was my only contact with the outer world. He didn't have much information, but in a clamshell, the fouled anchor was: the railroad had agreed to transport a box car with my horses, belongings and me from Loomis to Big Pine, okayed on both ends, but they had neglected to clear with Nevada, whose territory they were illegally trespassing. Nevadans bowed their necks.

"The sun dipped to the western horizon. Olfactory conditions had progressed. Various flies had selected my box car as a local, national and international convention site; their refreshment committees were energetic and efficient. Outlook was bleak.

"That night, I watched the tired lights of the railroad yard and the lamps of Reno glittered in the distance, to me only a mirage.

"The sun lumbered up with a negative look. My keeper brought water and victuals, but no news. He couldn't stay and talk; probably, the odor prodded his departure. Things got real bad that afternoon. I think the horses, too, were getting cabin fever. I decided that the next day I must take action. Bullets from the big gun of the big sheriff couldn't be worse.

"The sun was just above the western horizon when a switch engine puffed off onto my siding and, with no explanation, moved me back into the main railroad yards.

"After about an hour of pulling, shoving, jerking and banging, a long freight train moved east into the desert twilight with my car in the center.

"I wish I had a record of the debates that went on to decide my fate. Undoubtedly, some parts would make my ordeal sound like a Sunday school picnic.

"The movement of the train caused the cool night air to circulate through the open doors of the car as the gloaming changed to the blue black of a moonless desert night. The horses must have enjoyed the fresh air too because, except for when they were thrown about by the jolting and shocks of the car switching, they were leaning as far over their retaining stanchion all night. Suddenly, all motion stopped. I could hear the train's coughing gradually evaporate in the empty distance. A survey through the car door showed a few dim lights.

"Morning and the old mining town of Mina greeted me. The sun was just beginning its daily task of patiently returning the few houses on the one street and the Indian dwelling in the back to the desert. Everything was warped, splintered, leaning and weary. Not a smidgen of green. It was losing its battle: vicissitudes of rugged seasons.

"I wish I knew the history of Mina, then, its only excuse for existing it was the junction of the broad-gauge to Reno and the narrow-gauge through Owens Valley to Lone Pine.

"What the two lacked in its worship of the aesthetic, it made up for in its rugged warmth and hospitality, characteristic of Nevadans of that era. The people of Mina were not afraid of the hoof and mouth disease, they did not have anything more susceptible to hoof and mouth disease then a lizard.

"A tall, lanky, leathery, weather-beaten man, I expect major-domo of the railroad junction, took one look, or maybe smell, into the car and let out a string of oaths, *@*!@ *!@*!* what the h—l do we have here?? I explained my incarceration, and quarantine imposed by the Brass at Reno.

"With another string of impressive adjectives, he took charge. 'First, let's get those poor **!! *! horses out.' He arranged a ramp and we coaxed them down. 'I'll have the boys unload your belongings on the platform' then, addressing a group of young Indians who were admiring Cassy, 'You fellows take those horses out and let them roll, then feed and water them,' and to me 'Go take a bath and wash up and get a meal. There is a shower of sorts behind the restaurant.'

"A motherly, plump lady in the restaurant set out the most massive breakfast I have ever encountered built around an enormous steak. She was reluctant to take pay until I assured her that the government would reimburse me.

"The narrow-gauge wouldn't pull out until next morning so, to give the horses a little exercise, I rode them around the town followed by the juvenile population, about ten, mostly Indians. I asked them if they wanted a ride. They were hesitant, but after I coaxed one brave lad to get on, my day was made. With a couple on each horse and the rest tagging along behind, occasionally alternating riding and walking, we set out. As we gradually left the tin cans, bottles and discarded junk of the town, the desert was beautiful. It seems to me that when undefiled by man, nature has never painted an unharmonious picture. When we returned to town, somehow the secret leaked out that the horses would eat out of their hand and I imagined that some of the home commissaries were woefully depleted.

"That afternoon, the narrow-gauge had pulled in and 'spotted' a car at the platform. The 'Major-Domo' had our load packed in the car and braced and a bulwark to hold the horses on their side, which he installed when the horses were loaded the next morning.

"We clattered south across alkali flats, past barren hills and arroyos that even the sage refused to inhabit. Then down into the north end of Owens Valley. The old White Mountains to the east looked west with tolerance and patience at the robust upthrust of granite, the Sierras, so young that they still carried on their crests remnants of the ancient Calaveras formation that had pushed through. It would be a soul of iron who wouldn't thrill to the panorama when seeing it for the first time. It just choked me with emotion. I was hooked and surrendered to it without a struggle. It was my country and always would be.

"My car was stopped by a loading platform at Zurich, the railroad station about a mile east of Big Pine. The agent had me sign several forms and affidavits testifying to the world, in general, that the railroad had delivered me and my gear, more or less intact, to my destination, then lost all interest in me. I had been assured that there was a house for us to live in. Fine, but where!!!

"I had just unloaded the horses and was contemplating, with little enthusiasm, the job of unloading our possessions when a big man on a big horse with all of the equipment of an authentic ranger rider rode up. He took a big chew of tobacco, looked us over, especially Cassy, then 'Where the *@!! have you been? I have been looking for you for a week.' It was Bob Logan, who had been a ranger and now had a pack outfit on Big Pine Creek. He will be mentioned many times in the future. After I explained that I had been enjoying the hospitality of the Reno—Sparks Railroad Yards, Bob took over the operation. He called a man who had a truck and in a short time we were loaded and headed for Big Pine, Bob and I following with the horses. How exhilarating it felt to have Buck cakewalking under me.

"Our future home was a cute little white house with a bath, tiny bedroom, living room, ample kitchen, and corrals and a hay stack in the back separated from the house by a huge garden plot. Attached to the house was a miniature office, about 8' x 10'.

"A note on the desk asked me to call the Supervisor's office in Bishop. Supervisor Tom Jones answered. After 'garden run' introductions and salutations, he informed me that they had $3000 with which I was to build a trail over Bishop Pass, from Long Lake, which was the end of present trail, to King's River. He told me that Cranston, the Assistant Supervisor, would meet me in Big Pine in five days with my abney level, compass and a load of survey stakes.

"Cranston then came on the phone and asked me to tell Logan to hold two specific mules known for their proficiency on rough trails. Then he asked me if I had mules and when I answered no, he said that I had better hire some as horses would never make it over the trailless pass. I told him if my stock couldn't handle the job, it was a good time to find out.

"The horses had been in pasture all winter and their hooves grown long. No. 1 project was to get them shod. Problem No 1: There was no blacksmith!!! I was stunned. There was no other solution, I would have to do it. I had seen blacksmiths shoe horses and vaguely remembered seeing them applying the hot shoes to the hooves and seeing and smelling the smoke that emanated from the contact. That was my knowledge of the farrier's craft. Mr. Bigelow's words came back. 'A good ranger can do anything anyone else can do and do it better.' I did want to be a good ranger!! Listed in my inventory was farrier's tools which I dug out: hammer, rasp, nippers, and a few horseshoe nails.

"In the storeroom of the Big Pine Hardware Store, there were several barrels of 'cowboy' horseshoes and one barrel of steel ones used by blacksmiths. The cowboy shoes are of softer material and can be worked cold. The ones the blacksmith used had to be heated to be bent to fit.

"Being ignorant, I was confused, but the owner of the store, George Hall, (He, afterward, became a very close friend. He was Master of the Lodge and, afterward, gave me the third degree in Masonry.) was there to give me much good counsel. He sent me back to caliper the horses' hooves and picked out a dozen shoes to fit. He told me how much of the hoof to cut, or rasp off, how the nails should go and a lot of other good advice, some of which I remembered.

"Next morning early, I started on Number One of my hardest day's work: Cass was perfect, except her hooves were hard as flint. The nippers wouldn't touch them and they only yielded to the most persistent rasping. (The horses had been on soft ground pasture all winter, the hooves had grown long and had to be cut way back.) Fossil Face's feet were enormous and shaped like an oversized pancake. She was patient, but as soon as a foot was raised one would think that was the only leg to stand on. My back still aches when I think of holding those big feet up. I later solved that problem by putting a box under it to hold it up. Buck was perfect too, except he found it necessary at irregular intervals to give the foot a jerk. This was a real hazard. When the nail is driven through the shoe into the hoof, it is shaped to come out about 3/4 of an inch up the hoof and protrudes out about 1/2 of an inch at a rakish angle. The hoof is held between the thighs and braced up by the inverted knees so a devilish jerk of the horse's foot can well change the holder's voice from a hearty baritone to a girlish soprano.

"By sundown, I had it, but, for better or worse, I iron nailed on twelve hooves.

"In a couple of days, Cranston was there with his saddle and my allotted packsaddles and kayaks. Bob Logan had arrived with the mules that Cranston had requested, handsome, big black devils. We loaded our supplies and were off. Up to the head of Baker Creek, over the pass into Bishop Creek Watershed, up Bishop Creek to Long Lake where anything resembling a trail vanished. All of the time, Cranston in a tirade on the inadequacy of horses versus mules on mountain trails (generally speaking, he was right). He prophesied the gloomy tragedy that would result when my horses tried to go over 'the Pass.'

"Next morning, the going was good 'til we hit the moraines composed of rocks from the size of a bucket to the size of a piano. Working our way through the rocks was hard on the stock; then the steep upthrust to the summit. It looked impassable. Snowbanks covered what might have been passable routes. We knew stock had gone over it, but where? After trying several places, we found one cleft that looked passable. It was there where Cranston, who was in the lead, started to have trouble with his mules. His saddle mule refused to be led and in a short time both of the pack animals were laying down with their loads and refusing to get up. I knew that Buck and Cass had plenty of guts and would try as long as I asked them, but I didn't know about Fossil Face. The going was pretty bad, but not impossible. We unpacked the mules and unsaddled Cranston's mule. We got the horses around the downed mules and they scrambled, climbed and scratched to the summit. Fossil Face was no quitter!! We hadn't got out of sight when here came the unpacked mules following us. It wasn't that they couldn't do as well or better than the horses, it was just that somewhere in their cantankerous lives they had found that if the going got too tough, just lay down and the packs would be removed. We had to unpack my horses, return down that *!!@! route and bring up the jettisoned cargo. There were no more degrading remarks regarding my horses.

"We camped at the beautiful Dusy Lakes, took scattered, hopeful looks at routes where a trail might be built, then back to South Lake on Bishop Creek. Cranston had to be back in Bishop for something or other. He caught a ride back to Bishop. Cranston was a morose, efficient, hard working Norseman, an ex-plumber. We had not agreed too well on the trail routes. I was anxious to be back to Big Pine as I was expecting Helen in a few days so I took the rented mules and hightailed back via the Coyote range to headquarters.

"Helen arrived in a couple of days effervescing with excitement and smelling of sheep-dip that, somewhere in her pilgrimage, she had to walk in to liquidate any hoof and mouth bugs she might have on her shoes. Without taking a breath, she started changing our little house into a home, joyfully unpacking her furniture that she hadn't seen since we were married.

"Another first for me—this time to know what the term 'swamped with work' really meant!!

"The Coyote Range Association were due to drive their cattle onto the Coyote Range and the critters had to be counted. Paul Lucco's sheep were nibbling off, and swallowing with gusto, tidbits of Parker's Oak Creek cattle range and Parker was squalling over the phone like a cheated harlot. Across the Owens Valley, I could see the dust from two bands of nomadic sheep illegally slicing off generous hunks from the Inyo Mountain Range. Numerous letters from the R.O. demanding immediately the delinquent reports on Special Use Permits. Various requests for wood cutting from Independence to Big Pine. Local delegations offering unrequested advice on how the Forest Service should be run. Three bands of sheep to be counted into their ranges. No use to call for help as all of the Inyo N. F. personnel were having their own problems.

"In my ignorance, all was as tangled as a barrel of well shaken clothes hangers. Each night I poured over back mail, filed, and looked over the permits and maps trying to figure priority and coordinate the jobs with their location on the map. Counting cattle was not so bad, but, believe me, counting sheep does not put one to sleep, it drives one berserk. The poor horses really got a workout. I was working out of Big Pine and many nights I couldn't make it back, so I gave my horse a nose bag of oats and slept on my saddle blankets wherever I clonked out.

"The trail crew for the Bishop Pass trail were to be taken in, about a week. I had to locate a camp site for the crew and have the trail located and marked out. Helen was all excited about her first pack trip, planning menus, etc. . . .

"We started up Big Pine Creek making our first camp in the beautiful Big Pine Lakes Basin. The next day, over a ridge to Baker Creek, over another pass into Bishop Creek Watershed and then up Bishop Creek to Long Lake where we camped. Picking out a location for the crew's camp was easy, but when it came to the actual laying out and marking the route where the trail would be built, inexperienced as I was, it really looked hopeless.

"My equipment was an abney level, compass and the little Forest Service handbook on trail construction. The instructions were 'keep it at fifteen percent or lower.' (Now, there would be an engineer, two crew of fifteen with transit, plane table, drawing board, etc . . .). I had a swivel stick with marks the height of my eye and I also had a map level. We worked there for two or three days getting the trail on the permissible grade of less then fifteen percent.

"The trail crew arrived and to my relief the Supervisor's office agreed to take care of their time and provisions for the camp. The mountain Gods must have been in a generous mood because, among the crew, was an enormous Dane. This robust viking had the desire and the rare ability of a natural administrative leader. I selected him as Camp and Trail Boss which was unnecessary as he had already appointed himself to that position and, with rare good judgment, was laying out kitchen, bunk tents and latrines and within minutes had crews working on each project and a cook preparing lunch. He was such a help in laying out trails; what seemed like Herculean tasks to me, he would take care of this on his way to work. I remember watching him, with jealous admiration, as he, with a ponderous crowbar, moved across a granite moraine composed of massive rocks from the size of a wash tub to the size of a piano. He looked like a one man avalanche on a pilgrimage. I learned from him!!

"The trail route established and marked to the summit, Helen and I moved our camp to Dusy Lakes on the King's River Watershed. I always cherish the memory of that camp as it was then almost surrounded by 13,000 foot peaks and ranges, lots of pure water, blue grass up to the horse's knees and big, fat golden trout in the lake just begging to taste an artificial fly. Our one fly came unraveled so I manufactured my first one. Tail hair from Buck's breast, body from red yarn from Helen's sweater, hackle from Fossil Face's bay hair held together with thread unravelled from our trap and warm pitch. How those golden beauties loved it, to their misfortune.

"I was sweating blood trying to find a trail location that would qualify. The passage in use was a narrow slide between two cliffs which were too steep and too narrow for switch-backs. Things were looking pretty bleak when Tom Jones rode in with two blankets and a fishing pole.

"Supervisor Jones had the best mind I have ever known for analyzing and solving problems, if one could get his mind off of fishing. He would go to the office to work in the morning, instead of going home for lunch as Mrs. Jones expected, he would often hear of a ranger going on a two week trip, so he would call up the ranger to hire a horse for him. Remembering to take only his fishing outfit, he would be gone from home for a week or two. His wife would not know where he was or when he would be back, and had long since become used to it.

"Ten minutes after his salutation of 'How is fishing?' He had a big golden on his line. When he had caught enough to last us a couple of days and had caught and liberated a limit or two, we persuaded him to come in for a big meal Helen had prepared.

"The next morning, Mr. Jones' mind off fishing for the moment, we worked out a passable route along the steep bluff under the lakes where, to me, a trail seemed impossible. He gave us a stock of fishing flies he had made and headed back to Bishop leaving me to locate and mark the actual trail site and run the grades.

"Before Mr. Jones left, he told me that he would have $1,000.00 to build a trail to the top of Mt. Whitney, the southwest corner of my District. He said that if I was in that part of the district, to see if I could plan a trail that could be built for that amount.

"We decided that now was a good time to take care of that job. Both of us were anxious to go to the top of Mt. Whitney, which at that time was the highest mountain in the United States. We would go down the west side of the Sierras and come on to Mt. Whitney from that side.

"Oh, to have again that brassie stupid ignorance of youth!!

"We had hardly started when that fickle Madame Luck took a swipe at us. We were at a glacier polished granite slope just above where it plunged over a cliff down about 500 feet to King's River. For some reason, I was riding good old, big footed Fossil Face. With no warning her feet slipped out from under her and we were sprawled on that adamant granite. Thank the Gods of change, she fell up hill. Had she fallen down the slope, we would probably have had an exciting, but terminal, slide to the netherlands. I cleared the cantle of the saddle with my left foot, but my right foot was still in the stirrup and under Fossil Face when her ponderous weight hit the rock. I liberated my foot as she struggled to get her footing on the slick surface. I kept her head up hill while she struggled to her feet. When I got a break in Helen's vociferous broadside of questions, advice, and admonishing, I told her to dismount! Another lesson learned from experiment: When crossing slick, sloping rock surface, get off and walk! When I started to walk, I noticed a sharp pain between my little toe and heel on my right foot. Later found that the bone had been cracked. It gave me h—l for the rest of the trip and was very sensitive for a couple of years.

"After reconnoitering, we found a slide between the cliffs where apparently someone had taken stock down before. It didn't look good, but we could find no alternative. The last descent didn't take long. The horses literally sat on their fannies and slid along with a miniature avalanche they had created.

"King's River and beautiful meadows with wild flowers galore down the King's River Canyon where now is a portion of the John Muir Trail. The first few miles were a veritable Eden. The floor of the canyon was one or two hundred yards wide and footing was good, then we hit it rough as the gorge narrowed with cliffs, moraine jumbles of massive granite chunks. If we hadn't seen ancient camp fire sites where hardy mountaineers had camped, we would have considered it impassible. (When we went down the same route thirty years later on the wide John Muir Trail, we could hardly believe we had gone through.) With that I am a good writer, but I could not describe the beauty of that piece of King's River. About the time we thought that we could proceed no farther, we came to Simpson Meadows. Acres of verdant meadows with scattered yellow pines growing up the river.

"We made camp under a big pine at the river's edge.

"The horses' legs and knees were scarred and bleeding. Cassy had a deep gash on her shoulder where a big rock dislodged by my horse ahead had hit her, almost knocking Helen and her off of the trail. Buck had a loose shoe so we decided to stay there for a day.

"Ranger Simpson once said, after pondering on one of my yarns, 'Fishermen are all liars. Brainerd is such a good fisherman,' but one of my better fish stories occurred the next morning: the night before, after caring for the horses and setting up our primitive camp, I rigged up the fishing gear and quickly took enough trout to supplement our meager food supply, leaning the pole and line against the pine at the head of our bed when I was through. Next morning, I sat up in bed and could see the horses drowsing in the morning sun just below us. Somehow, I knocked the fishing pole over and, on picking it up, shook out a little line and flipped the fly into the water when it was taken by a fat, eight inch golden. Within minutes, without getting out of bed, I took all of those beauties for the day's menu. Try to duplicate that any place today!!

"From Simpson Meadows, it was easy going over a pass on the headwaters of Dougherty Creek and down Copper Creek to Zumwalt Meadows. Up Bubbs Creek where we made overnight camp.

"I cannot remember how we got from the head of Bubbs Creek to the pass, but Junction Pass was rough, real rough going. Not sure, but from the map, it must have been Foerster Pass. I can still see the horses gamely clawing their way up the last part. Over the summit and into the Kern River Watershed.

"From the pass, the ground was fairly level covered with idyllic swales and glades cut by sparkling little streams and dotted with miniature ponds and lakes. The rocky ridges, where they protruded through the meadows, were covered with dense stands of lodge pole pines. The crest of the Sierras framed a background to the east. We made camp by two little lakes, real gems, an emerald, velvet carpet, close to the southwest side of Mt. Whitney.

"Next morning, anticipating a hard climb on foot, we started early. The climb was disappointingly easy, only one place called the Chimnies did I have to use my lariat.

" 'Brethes there a man with soul so dead' who, when attaining the highest point in an area, be it housetop, flagpole, or mountain, is not inoculated with at least a small quiver of pleasure, superiority and pride to be able to look over and down on all earthly objects in view. The top of the United States!! (Glad we were there before Alaska bumped California.) Sorry that my fifth grade vocabulary is not able to adequately describe the view and our exuberance. Such a magnificent panorama in any direction.

"To the east a sheer drop of hundreds of feet from the foot of these cliffs, rugged canyons and ridges pitching steeply down for eight or nine hundred feet and terminating at a more gentle talus slope that flowed down to the

Owens Valley. Across the valley were the Inyo Mountains forming a west boundary of the Great American Desert. Beyond the Funeral Range, the Panaments, Death Valley and as far as the eye could see, the hills and ridges smoldering in the seemingly limitless vast expanse of the Armagosa.

"To the west, the more moderate slope of the Sierras leisurely descending to the San Joaquin Valley. To the north and south, the craggy crest of the Sierras with Mount Russell, University, Tyndall and other peaks all giving Mt. Whitney stiff competition for height.

"On the top was a bench mark that read 14,501 feet (now established at 14,495), also a one room stone shelter, at that time unusable as it was nearly full of ice. There was a register, as I remember in a tobacco can tucked in a rock cairn, showing very few visitors. Wish I could remember just how many. Time was short and I ran out a proposed trail by using my abney level and Helen's stadia rod mapping it on several pages of my notebook. We got back to camp really tuckered.

"We cooked a big pan of trout that I had caught the day before, and Helen mixed some biscuits while I took care of the horses. Oblivion 'til the sun touched the tops of the mountains the next morning.

"This day and the next are branded as two of our hard days. We got an early start as we had very little to pack. We must have crossed to the east side of the divide through Shepherd Pass. Way down below us was Owens Valley, so we knew where we were going, but how was the problem. If there was any sort of trail, we didn't find it. Down a steep, rocky canyon until we came to an impossible cliff, then scratch our way up the steep side of the canyon wall and contour around it to the ridge and follow it until its massive outcrops of rock barred our way, then back down into the bottom on whichever side seemed most friendly, then repeating the grueling progress again and again most of the day. Near sundown, we came to the semblance of an old road to a mine which we followed down to the talus slope where we headed for Independence, which we could see in the distance.

"We arrived in Independence after dark, tired and hungry.

"The only restaurant we could find was just closing, but the kindhearted owner made up a package of sandwiches and doughnuts which we ate while riding north. A couple of miles north of Independence a small stream crossed the road and went through a grove of trees a few hundred yards off the road. There we made camp. I threw the bedding off the pack and it had hardly hit the ground when Helen was on top of it sound asleep. I unsaddled the horses, gave them an ample nose bag of oats (horse feed had been so good on the trip we had held back on the grain, so plenty was left) and flopped down beside

Helen and in no time was in hibernation until Cassy's nickering for refreshments the next morning shook us out of our slumbers.

“I watered the horses, gave them the remainder of the oats. Dressing was no problem as we hadn't even removed our boots. By the time we had breakfasted on the remains of our generous supper the night before, packed and saddled, the sullen sun was glaring at us from the east and we were being paid an unwelcome visit by swarms of flies from Paul Lucco's nearby corrals.

“Before we had gone far, we knew that we were in for misery. Weeks in the cool atmosphere of the high elevations had softened our tolerance to heat and one night was not sufficient time to become acclimated to the Owens Valley summer temperature. Boy, oh boy, Owens Valley can get hot!! The sun that day could curl the spines of an asbestos horn toad. One brief, delicious stop where Sawmill Creek crossed the road, we washed and watered the horses in that icy stream, then for the last long haul. The saddles stuck to the horses, our pants stuck to the saddles and our skin stuck to our pants. There was little conversation, just endure as the horses plodded. All things end, at long last, the horses were unsaddled and rolling in the dust of the Big Pine corral and we were taking a long, heavenly bath. How good a bath can feel!!!

“Then there was the time that we got to the Tinnemaha Fruit Ranch and stopped there for a drink. Mr. Alcorn insisted that we stop and have lunch. Lunch was really dinner. We got rested up and went up to the Murrey Ranger Station. I think I cheated a day or two rest though.

“The most important job was under control (laying out the Bishop Pass Trail). We had circled the west half of my district, but reading the backlog of mail made me realize that some sort of organization plan was necessary.

“I broke the work down, as follows:

“Grazing: Field count of stock, checking brands, distribution, range utilization, trespass, salt distribution, water supplies, drift fences, and pastures.

“Trail Mountains and Construction: Survey of work needed, work crews, tools, powder and supplies, rerouting, need for signs, and checking pack stock.

“Special Use: Check applications for summer homes, water permits, water lines, pastures, laying out summer home sites, and Mineral Claims survey.

“Lands: Boundary survey and posting, and survey private lands.

“Recreation: Campground development, tables, latrines, garbage disposal, and tourist information.

“Fish and Game: Animal census, fish stocking, mountain sheep studies, and law enforcement.

“Fire: Warning posters, tool caches, and law enforcement.

Fires, the bane of most rangers' lives in the summer, was of little importance on the Inyo, called the asbestos forest. I found that fires could go off like an exploding firecracker, but usually, went out on their own accord. The Forest Service later got the *!!@ habit of drafting us and taking us to Southern California to fight the big fires, so we became hard to find.

"Public Relations: This was stressed in those days of the Forest Service. The policy is probably illustrated by a pep talk given to me by supervisor Bigelow. 'Brainerd, you can't make everyone like you, but I demand that you make everyone respect you.' Damned and cursed as the Forest Service often were, they were always respected.

"The men and their wives of Inyo had an additional problem. The water war with Los Angeles was on in all of its fury, and the population was sharply divided with no blending. One was 'for or against' PERIOD. We followed the Tom Jones' suggestion: 'Listen politely to either side and make no comments.'

"I outlined trips on the map and listed in my notebook the jobs to be done on each trip with notation on their priority. With many improvements, this system became my annual work plan.

"Early in our stay in Inyo, we became acquainted with the Alcorns, later to be our lifelong friends. Harry and Delpha, the father and mother, have long since joined their ancestors, but we still correspond and have the warmest friendship with the children, about our age.

"Mr. Alcorn was the Superintendent of the Red Mountain Ranch that ran both sheep and cattle on the forest. I was not present at the first meeting, but it must have been interesting. Mr. Alcorn called and after introduction said that he wanted to see me to get a permit to fence a meadow for a buck pasture. Now, a ram Helen probably knew, but what she did not know was that in nomenclature of the range country, a mister sheep was called a buck. Her interpretation of a buck was a papa deer. A fenced in mountain meadow filled with massive deer, their antlers gleaming in the sun presented to her, an impressive picture. I am not sure just how the conversation progressed, but she entered into the project with her usual overabundant enthusiasm. Enticing the deer into the corral and what we would do with them were minor details that could be left to Mr. Alcorn and me. She assured Mr. Alcorn that she was sure I would cooperate to the fullest.

"When I got home that night, Helen pounced on me all agog with the operation. After listening to her 'til she could get her breath, my thought was 'suffering sidewinders, my largest permittee has eaten too much locoweed.'

"The buck pasture deal straightened and explained to Helen's disappointment, we were adopted into the Alcorn family.

"It was not long before I had to go to the Bishop Pass Trail operations. It was on this trip that Helen had to take the wilderness on her shoulder alone. We were camped by a little meadow by Long Lake. Helen stayed in camp with Pickles, a little black cocker, while I rode down to the road-head at South Lake to check on delayed powder and supplies for the trail crew.

"I had hardly swung out of the saddle when a telephone call came through for me to get to the south end of my district as there was a fire near Independence. I wrote a short note to Helen and sent it with my six-shooter and horse back to camp by a packer. I was gone four days and Helen was alone with the horses and Pickles, rather confused where I was and when or if I would get back. Our camp was about 11,000' elevation and miles from anyone. She handled the situation like a pioneer wife, caring for the horses and camp, catching a few trout, and collecting and pressing wild flowers. Except for a little worry, she had a wonderful time.

"The summer was a great adventure. To keep from backtracking, we worked two long trips: Over Taboose Pass, one of the toughest, then down what is now part of the Muir Trail to the Woods Lakes Basin and back over Sawmill Pass to McMurry R.S. The other one was over Sawmill Pass to the south past Rae Lakes back to the east slope via Kearsarge Pass to Independence, across Owens Valley to the Inyo range, up Mazourka Canyon to Squaw Springs and back to Big Pine by the Waucoba Road.

"We worked 10 or 12 hours a day, sometimes much longer. We almost always stayed in one place on Sundays: wrote up notes, washed clothes, cared for the horses' hooves' shoes, cooked a couple of day's food. Sometimes, we would establish a base camp and work out from it for several days. Helen took to fly fishing in a big way. Her quick reaction would hook twice as many strikes as I could.

"One day when I came into camp, Helen was all a'dither. She had been fishing in Sawmill Lake. She was about to start back to camp with a string of plump rainbows when suddenly, there was a loud roaring ssshwuish and the lake came churning up around her feet. She made a sprint, hit the saddle and was halfway to camp before poor Cassy was fully awake. Even after she had calmed down enough to describe what had happened, it didn't make sense.

"When we went back to the lake to retrieve her fish and fishing pole, the enigma unraveled. There was a large snowbank in a steep gully just above and extending down into the lake. As the support for the snowbank melted, gravity took over and hundreds of tons of snow and slush were dumped into the lake. No wonder that Helen, whose nature is to concentrate on effect rather than cause, had her stitches unraveled.

"Our Bishop Pass funds ran out just when the trail reached the last 500 foot pitch into King's River. I am still proud of the amount of trail built for the money. Much credit goes to the big Dane foreman. Wish I could remember his name.

"I came home one evening in September, after a long day's ride. Helen had a big dinner ready and the look on her face was like a fox terrier that had just killed a couple of the neighbor's prize pullets. After a little solicitous greetings, she suddenly blurted out, 'I got a job!!'

" 'Yeah, what doing?' I thought maybe making curtains or a dress for one of the neighbors.

" 'Teaching school!!'

"I wouldn't have been more surprised if she had said, 'Shoveling coal for Satan.'

"A prelude: When Helen graduated from high school, she planned to go to college. One of her friends, a classmate, was going to Normal. Helen tried to persuade her to go to college and in the ensuing argument, bet that she could pass the teachers examination with no further education. She 'boned up' and when the next teacher's examination was given, passed it easily, then forgot it and went on through college.

"While I was gone that fateful day, Ethel McAffee, a fifth grader from across the street, came over to visit Helen. In the conversation, Helen asked why she wasn't in school. 'We don't have school, the kids threw ink bottles at the teacher and she quit.'

"Helen remarked jokingly, 'I passed the teacher's examination. Maybe I could teach you.' Ethel had hardly time to report when Mrs. McAffee, who was a trustee, came bounding over. Helen tried to explain that she had just passed the examination five years before and had no other credentials to qualify her to teach. Ignoring such trivial technicalities, Mrs. McAffee at once rounded up the other trustees. There was one of the Shveart clan, big sheep owners, then big Alee McDonald whose enormous, sledge hammer fists could subdue any man in Inyo County. George Hall of the hardware store was Clerk of the Board.

"Helen tried to explain that she had no certificate and could not legally teach. Helen didn't have a chance. A teacher was desperately needed, she had passed the teacher's examination, legality was of little or no concern. She was a schoolmarm and that was that.

"It would be hard to find one as ill prepared to teach a frontier grammar school. Her schooling, up to her second year in high school, had been in convents in Mexico. She had never been in a grammar school in the United States, none of her college courses had touched on education. She was given the fifth and sixth grades.

"That year, the Indian school had been integrated with the white. After the manner of the Indian Service, the Indian school had to be run, as follows: the boys raised a big garden for the Indian Agent the produce of which he sold for his own profit. He was teaching them agriculture. The girls learned domestic science by doing housework for the Agent's wife. As an extra bonus to Helen, all of the Indian children were put in her class. Everyone from just out of the papoose stage to big booted and spurred lads of 18 years.

"Events climaxed fast. About the second day, a big, supposedly incorrigible lad was walking around the room in class. Helen ordered him to sit down. He approached her grinning and said, 'I would like to see a little duck like you make me.' I guess for the first and last time in her life, Helen struck a fellow being, but it must have been a good one. She really saw red and took an unprofessional, but efficient, wide arm swing connecting on his mouth and nose. To make it more effective, his nose spouted blood. From then on, she had control of the class.

"When the first report cards were given out, a boy from a prominent family got low, low marks. His excuse to his parents was that Helen wouldn't show them how to do the lessons. The parents complained to the trustees, so there was an investigation committee composed of McDonald, Mrs. McAffee, George Hall and the boy's parents. They sat through the lessons. All of the pupils, except the boy seemed to have a good understanding of the questions. Alee McDonald turned to the parent and summarized the investigation in these few words: 'Bill, your d—n kid is just dumb.'

"After that, the sailing was smoother. The Indian children were not the problem that was expected. Even though scholastically far behind, with proper instruction, they did very well, some of them giving the whites stiff competition, especially in written work. It was hard for them to recite. They were handicapped by their habit of dropping out of school for a week or two to go on jackrabbit hunts, pine nutting or powwows. Regular attendance, they considered a white man's weakness.

"Helen was a natural teacher. Her enthusiasm to learn was contagious and she had the ability to inspire the students. She expected college work and got it. In after years, many who attended her classes said that they had never worked so hard in their lives.

"When she resigned because Naeda was on the way, she received a letter from the County Superintendent of Schools praising her work and saying that any time she wanted, there would always be a teaching position for her in Inyo county.

"Helen taught three years in Inyo, remedial reading in Tule Lake and home teaching of Naeda in the Lava Beds. It was not until 1939 that Helen went to

U. of C., took the necessary educational courses, and qualified to legally teach under the laws of the State of California.

"As fall progressed, there would be no more long pack trips for the season. I could cut down expenses by putting Buck and Fossil Face out to pasture and just keep Cassy for riding. Most of my work could be done in a one day's ride from home. With just one horse, the problem was how could Helen go with me on weekends? The problem was solved by the purchase of an old cart and harness. Question: How would Cass respond to dragging a contraption around behind her? She had always been a saddle horse. I snubbed her to the post and, while Helen fed her tidbits, I put the harness on, pulled the cart up 'til the shaves were on both sides. I then attached them to the harness and attached the tugs to the single tree, got a good grip on the reins and climbed into the seat. Helen untied her from the hitching post and I gave Cass the word 'get up' and waited for the explosion. To my surprise, and maybe a little disappointment, she moved off as if she had been pulling a trotting sulky all of her life. (Later, when I harnessed Buck, it was a different deal.)

"What memorable times we had with that old cart. Almost every weekend we would go some place. Along the Owens River fishing, quail or duck hunting, or up in the foothills for cottontails or mountain quail.

"The cart had a liberal rack under the seat and often we would load it with hay, grain, bedding and go for a two day trip up in the snow and make camp. Sometimes far back in the Inyo mountains, always planning to bring back enough game for the week.

"Cass was a beautiful trotter, always carrying her head like an Arabian stallion. I guess people thought we were a wee loco, at least they always gawked at us with their mouths open, especially when Hogan, our dog, was poised on Cass's back where he liked to ride if it wasn't too cold.

"On Helen's Christmas Vacation in 1925, we took our first trip to Los Angeles. Our big mistake was to call up some of Helen's relatives, all God fearing people, but so hospitable. They must have vowed that they would entertain us every second and they did. Their idea of entertainment was to show us to other relatives who saw to it that we looked at more distant relatives. The only relief from relatives was to take us to Seventh-Day Adventist Churches. We did have one grand dinner at Casa Verduga sponsored by Helen's father's ex-wife. A valuable lesson well learned was when on vacation and have relatives in the area that you must call, do it just 1/2 hour before your transportation leaves.

"I bought Helen a pearl necklace for Christmas on that trip and we spent the rest of our cash on a set of Haviland dishes on sale. I remember we were so broke that we had to split a cup of coffee on the trail going home. Note: The Haviland of which Helen was so proud were all stacked in a dish pan for washing met a shattering end when the pan slipped from Helen's hand and landed on the floor. The only survivor was a platter which, somehow, was not in the dish pan. We still have and cherish that platter.

"As mentioned before, the Los Angeles-Owens Valley Water War was on in full blast. What was right and what was wrong, like most questions, depended on whether it was taking fodder out of one's stomach or putting it in. One tragic result was that farmers who had probably never been out of debt and now finding they were in possession of $10,000 to $40,000, considered themselves business men eager to invest and were ready-made targets for fast talking Los Angeles shysters and ended up broke. Ranchers who had never been out of debt and being able to sell their ranches for twice the amount. A person had to be neutral. It was like a skunk on one side and a spoiled cat on the other side. Arguing which one stunk the most. There were some very good people on both sides and stinkers on both sides, as there always is. I knew I was neutral as far as the water was concerned, but it was hard to prove that you were neutral.

"Methods of separating sellers were varied. I will narrate one where we knew the fleecer and the fleeced:

"S- owned a good ranch and water rights and sold it for $80,000. A Beau Brummell S- was not. S- owned and herded his own sheep. S- was a very good friend of ours and was respected in the community for his honest and sterling qualities, but, if anyone had ever told him of the benefits of washing, shaving or changing clothes, those words fell on alkaline ground and never sprouted. A wag said that if you saw S- going down the street and the shirt tail was in, it was not S-.

"A woman in the town whose background should not be looked into sized up the layout and made plans. Her sister, a snazzy looking matron who had not been a recent visitor to the area, suddenly showed up. S- was immediately blitzed with dining invitations, married the sister and was taken to Los Angeles. Then, before S- could gain his equilibrium, he was swarmed over by lawyers and a suit for divorce (one complaint was that when in a restaurant he picked up his steak and ate it with his fingers).

"A bewildered S- was back in Big Pine in a few days. He sought out a close friend of his and ours, a man well versed in legal matters, and poured out his dilemma and said, 'George, I am giving everything I have left to you.' About $40,000. The other half had been invested in an auto, a house and other things, all in the wife's name. S- had signed them all over to her with no idea of what he was signing. After the furor was over, George invested what remained for S-, so that he could live the way he wanted to as long as he lived.

"Later, at a bull session at Glacier Lodge, the following conversation: Bob Logan: 'How much did she beat you out of?' S-: '$40,000.' Bob: 'How many nights did you sleep with her?' S-: 'Two.' Bob: 'Twenty thousand a night. Was it worth it?' Big Alee McDonald: 'Anyone that sleeps two nights with S- has earned every d—n cent they get!!'

"The business people, seeing their customers leave and their means of support swindling, were, naturally, desperate. The local bank having their high interest paying loans paid and afterward proven to be in desperate financial condition from poor mining investments, were the main instigators of the war against Los Angeles.

"The aqueduct and water system to Los Angeles was blown up at regular intervals. Even the children were in training for what we would now call gorilla activities. When I was a lad, on the 4th of July, good entertainment was to put a firecracker under a can and pop it into the air. In Big Pine, firecrackers were used for more serious work. The kids would build a miniature Los Angeles aqueduct then, at strategic points, using their firecrackers, would blow it to h—l.

"We became innocently involved in some beautiful, successful mayhem: coming from trips in the high Sierras, we usually planned to come home late riding through the foothills and hot, talus slope in the cool of the evening or night. This night, we stopped where an irrigation ditch ran under the comstock that delivered water to the Big Pine power house. We watered the horses, rinsed our hands and face and rode on the few miles to the Big Pine Ranger Station. We unsaddled and fed the horses, ate a bite, took a much needed bath and surrendered to Morpheus until late morning.

"That night, some of the local lads decided that it would be a good idea to give the city of Los Angeles an annoying jab. Their project consisted of touching off dynamite under the comstock at the location where we had watered the horses.

"What a beautiful target we made for the Los Angeles Sherlocks who swarmed over the area. To strengthen their hands, while we were on our trip Tom Jones had delivered our quota of explosives, a couple of cases of giant powder with caps and fuse. He had cached them in an abandoned 'Chick Sales' in the back yard. (Can anyone imagine leaving two, fifty pound cases of powder with cap and fuse in an unlocked privy these days??!)

"I had previously borrowed a box of caps, a roll of fuse and about a dozen sticks of giant powder from the hardware store. The next morning, when I went down for our mail, I took the powder, caps and fuse to repay the hardware store what we had borrowed. A friend of ours, days later, told us that there were 20 Los Angeles 'dicks' shadowing me to town and back. Guess the evidence against me was so obvious that the rubber heels suspected a trap so didn't descend on me. I don't know how they decided that I wasn't the bomber, but I never knew that I was such a prime suspect until days later. Cases against the dynamiters were hard to get a conviction as it was almost impossible to get a jury where at least half of the jury were not friends or relatives of those on trial.

"Through the first Summer and Fall, although still quite primitive, our living had become more streamlined and efficient.

"Helen soon became quite proficient in her commissary job. Meals were planned for the whole trip. If bisquick or pancake flour were in the market at that time, we hadn't heard of them. Personally, I think that Helen was the one that invented Bisquick. She would prepare a mixture of flour, baking powder, salt and a little sugar and with a little bacon grease added was the basis for our bread and pancakes. We would fry the mixture or put it over coals. A little melted down sugar with a smidgen of maple sugar added was our syrup. One of my favorite dishes was noodles, chipped beef, onions, cheese and tomatoes. Of course, coffee was imperative. Bread was usually made in an Army mess pan with a closed cover and cooked in the coals. Bacon was always with us.

"Helen usually included a few pints of concentrated meat for occasional stews. As mentioned before, trout were our main source of protein. Helen could keep us supplied with them while I was working. Anything canned was a luxury. We did take canned milk, tomatoes, small cans of crushed pineapple to freshen our dried fruit—peaches, apples, apricots and prunes. When sage-hen and grouse season opened, we really lived in luxury. Once in a while we would have young groundhogs to eat. Only the young groundhogs are good. Porcupine is alright if it is young, so we had it quite often.

"Often we would finish the day by reading and drinking a tea made of a high mountain mint. Just before going to bed, we would put dried fruit, usually prunes, in our coffee pot, add the proper amount of water and bury it in our camp fire coals. They would be well cooked and warm for our morning fruit.

"We tried dehydrated eggs and milk, but the product of those days both carried the aroma of rotten eggs and tasted the same so they were scratched off of our menu.

"The first two years, we did not have a tent and in case of rain made a lean-to by throwing our tarp over our pack rope stretched between two trees.

"Our bedding consisted of three army blankets, a quilt and a 5' by 12' tarp with grommets on the sides so it could be tied up like a sleeping bag. Our mattresses were as varied as the material available. Sagebrush tops, or various grass and weeds were the poorest; boughs of pine, cedar, juniper with fir boughs the best. In the absence of other things, we used the saddle blankets and pads. They were better than nothing. We tried to rinse them out when we had time which was not often so when using them, we often crawled out of bed impregnated with a high, gamey, horsy aroma clinging to us.

"To me, there is still nothing better than a well made bough bed: two foot branches were laid overlapping each other, then one foot pieces were placed on top at a steeper angle and for real luxury another layer of finer twigs were placed on top, almost upright. I can still smell those wonderful beds. We used our tucker bags with our extra clothes for pillows.

"We had one little trick that helped take the curse out of getting out of bed on cold mornings. We would lay a pile of inflammable material about six feet from the head of our bed, then place a stick with grass, paper or pitch shavings tied to the end and put it under our bed. On the frosty morning when it was time to get up, we would just take out the stick, light the end, reach over with it and light our bonfire and, when it was burning well, we would crawl out of bed and dress by it. Sure felt good on one's hind legs on cold mornings and they could be cold! I remember when Helen's dad was camping with us. He had put his uppers and lowers in a cup of water. When he took them out in the morning, they were grinning at him from a solid block of ice. I remember his comment, 'I thought it was cold in South Dakota, but I never heard of a man's teeth freezing together.'

"In early spring, I brought Buck and Fossil Face in from pasture, shod and grain fed them and rode them hard every day getting them toughened for summer work.

"Bob Logan was going to the McMurry range on the Inyo Mountains to round up and bring in his saddle and pack stock that he had wintered there. I volunteered to go with him as I wanted to check the feed and water holes on that range. He was glad to have me as the Val Dias brothers had not shown up yet and a rider with tough, well shod horses was very welcome.

"I had rounded up cattle, sheep, sheepherders, lost tourists, etc., but corralling horses and mules that had been free on a good bunchgrass range was something new. Never have I had such a wild, hazardous ride. Bob was a superb rider and roper. His instructions were to try to cut out or rope the lead mare. There were about 30 horses and mules. He sat on his big chestnut gelding like he was part of the horse. He would plunge down canyons and over cliffs at a full gallop, places I would usually go down at a slow walk. As I got excited in the chase, and with Bob as an example, I found myself riding as I never thought I would do in my wildest nightmare. I had a couple of chances to make a throw at the lead mare, but I was always a ham with a lariat and only ended with my noose encircling a bunch of sagebrush. After about three hours, we got them on more level talus slope where they were no match for Bob's big chestnut and he soon had his rope on the lead mare. I took my saddle off of Cass and put it on Buck who had been faithfully following. From then on, it was easy. Bob took the mare down to the road and I swung around behind the adamant ones and soon they were obediently following Bob and the mare. By the time we got to Bob's corral outside of Big Pine, the stock had changed from a scramble of fiery mustangs to a string of docile pack animals—all except one!! A mouse colored little jenny that was acting like a coyote that was high on locoweed, trying to get away, but wouldn't leave the bunch.

"When well corralled, Bob looked his stock over and said, 'How the *!!@ did you ring in that jackrabbit?' He said that it was not his and, as we could see no brand on it, concluded that it was a wild one. Bob rolled a Bull Durham fag and with a devilish look in his eye said, 'Could you use another pack mule?' I could. He roped the little stinker, snubbed her up to his saddle horn, put a heavy halter and lead rope on her and said, 'She is all yours.' I was flabbergasted!! Yes, I wanted a pack mule, but not now. All I wanted to do was to go home, take a bath, eat a big dinner and stretch out, but, after the good show I thought I made that day, I couldn't go 'chicken' in front of Bob now, so I changed the saddle from Buck back to Cass, took a couple of turns of the lead rope around my saddle horn and started home. I had to go through the main part of Big Pine and what grand entertainment we must have provided for the residences of that pueblo. The little misbegotten daughter of Satan was under Cass, in front of her, over her neck and on her back behind me. Good old Cass plodded on with only one thing on her mind, the nose bag of rolled barley she knew she would get at home. Buck followed along like a faithful dog. What a relief when they were all in the corral with their well-earned grain and hay. Thus, 'Cactus' became one of our family.

"Supervisor Tom Jones was transferred to the Shasta (We will miss his brilliant mind and his instructions on taking trout), Roy Boothe from the Sierra took his place. Roy will always be my ideal of what the public servant should be. He had been a ranger and District Ranger on the Sierra National Forest since young manhood. He must have been in his late 40's when promoted to be Supervisor of the Inyo, tall and handsome and dedicated to the idea that the National Forests were created for the greatest good of the most people for the longest time. I was attracted to him when I was going to Ranger

School on the Sierra in 1922 and was delighted to know that he was the one who would replace Tom Jones.

"It seemed to us that the first results were binding the rangers of the Inyo and their families into one large, cooperative family. During the winter's slack season, we would work as a unit in different districts on various projects.

"We soon felt the effect of his administration on the Big Pine District: after the Indian children had been integrated into the county schools, we had the Indian Agency property, big school house, residence and large lot to the Forest Service. The residence was a very well-built, atrocious dwelling: high ceilings, few, small windows, many small rooms painted inside and out a dysentery, gangrenous, dark yellow. Helen took one look, 'I won't live in it!' Roy Boothe gave it a look, 'I don't blame you,' then, 'What can we do to make it livable?' We had a meeting and appointed Helen as architect and designer, Pat McMurphy, who seemed the most qualified, as foreman, and Simpson, Boothe, Way and I as workmen.

"Under Helen's direction, and using material scrounged from the school building, we put in a false ceiling about 2 1/2 feet lower than the original one, installed 10 extra windows, rewired for electric plugs and for side lights. (The originals were droplights from the ceiling, standard in Big Pine at that time.) We rebuilt the kitchen, changing the sink from a dark corner to a cheery place under a window and put in ample drainboards and working space on each side. From bookcases from the school, we fabricated plenty of cabinet space. Remember, we had no funds allocated for this project. Paint?? We had plenty of war surplus, but it came in two colors—white and dark—battleship grey. Leave it to Helen, Under her administration, we made a light grey by mixing white with some of the battleship grey for the outside, then trimmed with dark grey. With the various sign paints the forest Service had left over, red, blue and yellow, and using the surplus white as base, she concocted blends so the whole interior had a bright, cheery, warm glow. The finished house was the envy of Big Pine (There wasn't much competition) and the Regional Office was flabbergasted. There was no costs, except our labor. It was a show place of the Forest Service.

"The spring of 1927 was well along; I was breaking my arm patting myself on the back that I had things pretty well under control and could keep them that way. I had just returned home from helping Bob Logan bring in his saddle and pack stock from the Inyo. This time with his stock was a beautiful wild sorrel mare. Did I want it? I couldn't resist. The journey from his corrals home was a repetition of my trip with Cactus, except that Winnaduma, which we named the little sorrel beauty, fought, bucked, plunged and kicked at the end of the leash rope rather than trying to ride in front of me, behind my saddle, etc., as Cactus did.

"I had just gotten squared in front of one of Helen's delicious dinners when the phone rang and it was Roy Boothe's soft voice. 'Have you got your work pretty well under control?' 'Yeah.' 'Good, then, we have a little job for you. Holetting is being transferred to Trinity, Guy Way is going from the White Mountain District to the Mt. Whitney and, as the Region has no other man available that they think could qualify, I told them that you could handle it in addition to the Big Pine district.' As I was too stunned to reply, he added gently: 'Your pay will be $1800.00 a year.' I had recently had a salary increase to $1560 and a $300 a month raise at that time was fabulous.

"I already had the east slope of the Sierras from Bishop Creek to Mt. Whitney and the Inyos from south of Independence to Westgard Pass. Adding to this the vast White Mountain District, which took in the White Mountains from Westgard to Montgomery Pass, then West, the Pizzoni Range and Benton back to Bishop.

"It wasn't a question of if I wanted it, I had it!!

"I did not want to take a trip through my new acquisition until Helen was through teaching and could go with me in about three weeks.

"After corralling and completing all the foreseeable jobs that should be attended to on the Big Pine District, I transferred the White Mountain files to Big Pine. Then, pouring over these files and maps, with Simpson's help, we planned a trip covering the entire district listing grazing permittees and number of stock allotted, private lands, springs and water supplies, special use permits, wood sales requested, allotting time for unforeseen things that would come up. We figured a month for the trip; if I remember right, it took 33 days and no place to replenish our supplies, but, by now, we were well prepared. Cactus had become a dandy pack animal and Winnaduma was partly broken giving us a strong of five, all of which could be used for riding or packing. They were all grain fed, hard and well shod.

"Another adventure into a new country, Helen and I on Buck and Cass, Fossil Face and Cactus carrying provisions and gear and Winnaduma loaded with grain, Hogan and Pickles ranging ahead as vanguard clearing the road of lizards and chipmunks and hazing them home to their burrows.

"We followed the Westgard Pass road to the Toll House, then north and east to the summit and made camp in a little meadow near the ancient, hoary Bristle Cone Pines. Next day, leaving camp, we made a trip up among the Bristle Cones and back to camp checking the range and springs.

"I so well remember cutting a limb of a Bristle Cone and marveling at the microscopic annual rings and commenting on how old they must be. Not realizing that in years to come, they would prove to be the oldest living thing known.

"The following day, we broke camp and rode Wyman Creek to the range cabin and contacted the Galloping Swede, range foreman of the Deep Springs. I think his name was Ranger. Bob Logan once found him vainly trying to move a herd of cattle, charging back and forth, with no idea of how they should be strung out and moving. Bob said to his companion, 'Would you look at that Galloping Swede?' The name stuck. He came to Deep Springs with a degree in stock raising from a Swedish college and had a lot of trouble adjusting his European education to western range management. He was a brilliant man and with his college foundation, eventually became a wonderful stockman.

"We spent a day with him, then to Crooked Creek.

"Paul Lucco had his range north of Crooked Creek and Lasague ranged south of the creek. Lucco and Lasague were both typical, old time sheep men. I had several run-ins with Lucco for shaving generous slices of range that were not his from National Forest land. He explained it to me, I wish I could put it in his words, but it boiled down to this: Raising sheep was like gambling. If everyone played square, by good playing, one might come out ahead, but, if everyone but you cheated, you lost. With handling sheep everyone stole range, so he had to. Paul Lucco had no intentions of going broke.

"We soon found sheep tracks. We followed them where they had grazed for three days, indicated by where they had made night camp and at last came onto sheep, with a herder and two dogs in charge. I proceeded to inform him, in official terms, that the sheep were in trespass and anyone that would graze on another's range was a bit of a cad. It didn't register, so Helen tried him in French and Spanish. He shook his head and played the sheep herder's trump card when they think that non-communication is the best policy. Tapping himself on the chest and shaking his head said the one word, 'Basque.' They know darned well that no one, except another Basque, should know their weird language. One might as well try to understand Finnish. Trying to get satisfaction from him, I knew, was like a billy goat butting a stone wall, so we dropped down to the main trail and headed down Crooked Creek intending to make camp at the Crooked Creek Ranger Station.

"About a mile from the Ranger Station, in a beautiful little cove, we located Lucco's base camp. The following will always stick in my memory:

"Lucco was just finishing making sheepherder sourdough bread (how good it did smell). He wiped the surprise off of his face in a second and came to us with a smile on his face showing his big, white teeth under his huge, black handlebar mustache. 'How glad I am to see you, I was expecting you.'

He told us, 'My friends get down, get down.'

'I am not going to get down, you're sheep are on the other side of the range,' I said.

Then he said, 'Oh, surely they got mixed up.'

I said, 'They didn't get mixed up.'

'Oh, no, I didn't tell them, but they do get mixed up,' He said.

I lowered the boom on them and everything I could think of, dire threat and such. I got through talking to them, he said, 'Ah, now my friend get down and we have some soup.'

'I don't want any of your d—n soup,' I informed him.

He said, 'You're not eating any of my soup then you are a friend no more.'

We were both hungary as dukes.

"The Basques use the sheep's head in their soup sometimes, at first, it is a shock to see half a jaw with teeth grinning out of the bowl at one, or poached sheep's eye glaring at you from your spoon.

"Another lesson for me: Whatever provocation, bitter quarrels, disagreements, one must never turn down his hospitality at meal times.

"Using the Crooked Creek Ranger Station as our base camp, we worked the White Mountain Range from the summit to Fish Lake Valley, the headwaters of Crooked Creek and McAffee Creek, almost to White Mountain peak. Sure enough, I found the Gireau and Lasague sheep on McCord range 5 miles from their own having shaved off a generous west part of Moline cattle range in transit.

"I knew that trying to get a federal case against those rascals was not practical and, unfortunately, they knew it too. I got them all together at Lucco's camp where they all swore, wishing that their mother's soul was burning in hell if they lied, that they didn't know where their range boundary was, for certain, and were unaware that they were off of it. (They knew every section corner, spring and badger hole.) According to their statements, they were all as innocent as unborn babes and any infringement on another man's grass was an unavoidable mistake. Alfred Gireau observing, 'We must never get on cow man's land because they get cross and do things to sheepherders.'

"I laid down the ultimatum, 'Anyone found trespassing would have to sacrifice twice as much of his range to those who had been trespassed on.'

"As we could only give the range an inspection once or twice a year and, as they were all three master range poachers, they should come out even. From a sense of duty, they would steal a little cattle range, but they would be careful because 'cattlemen did things to sheepherders.'

"After reluctantly leaving the Crooked Creek Ranger Station with its good pasture and abundant horse feed, we worked our way up through the Moline and McCord ranges and established another base camp on the headwaters of Chadovich Creek.

"From there, we covered the northeast, north and northwest side of White Mountain Peak, at times making overnight spike camps. We inventoried the mountain sheep on the northwest slope of White Mountain Peak. Made an estimate of wild horses on Kennedy flats and plotted springs on our map. It was another extra nice camp and of much interest to us. Many of the huge boulders were covered with petroglyphs. Although all were pre-columbian, some were sharp and clear, others were so ancient and eroded that it was necessary to run a chalky rock through them to make out the characters. Although there must have been a vast difference in age, the same markings were found on both. Professor Douglas, University of California, established the age of the Bristlecone Pine at about 4200 years old. He took peoples reports on the Bristlecone. Professor Douglas went up Shotovist Canyon and made base camp at the head of the Creek. There was lots of gold and petroglyphs. All of them were old, but some of the pictures must of been two hundred to three hundred years old. Some were much more dim then others. The day before making camp here, while crossing a deep arroyo on the bottom of which a huge boulder had been exposed by a recent freshet, the exposed part of the boulder was grooved with the same marks and symbols as those near our camp.

"After breaking camp, we crossed the summit of the White Mountain Range south of Mt. Montgomery and dropped down the west side of the mountains to the shimmering heat of the valley crossing to Benton, then up into the Pizzoni Range. Of much interest on this range were the numerous circles of stone where pre-historic Indians had camped. These numerous camps must have been for pine nutting as there were large groves of Pinon trees nearby. A secondary occupation while these camps were occupied must have been the making of obsidian objects inside these stone circles. Adjacent to them were great masses of broken arrowheads and scraping knives in all stages of construction and enormous quantities of obsidian chips. I imagine they manufactured them for winter.

"We had been gone a month and our food was getting low and weary. Those beautiful trout that we got so tired of in the Sierras, how good they would have tasted. There were lots of sage hens, grouse and deer, but the season was closed. About the only fresh food was occasionally wild onion.

"We headed south, through the Indian Reservation in Bishop, flow after flow of basaltic lava so barren that even a jackrabbit would have to carry a lunch and canteen.

"We stopped the last night in Simpson's house in Bishop and had a good bath and the best dinners we could buy in Bishop. The first thing we did when we got to Bishop was eat three big combination salads. We took lots of dried fruits—peaches, apricots, prunes and raisins. Quite a bit of macaroni, rice, and beans. Beans took to long to cook in the higher elevation, so we had to cook them at the base camp. We all took bacon, but at that time we did not have to worry about calories and cholesterol. We had a jar that we collected bacon grease. I would shoot Jackrabbits and we used to eat an awful lot of trout.

''Next day, down the hot valley to Big Pine.

''Then, as now, gigantic fires, fanned by the devil blessed Santa Ana Winds, raged through the chaparral clad mountains of the southern California forests. Overhead personnel on these fires was always at a premium. Somehow, the Regional Office got saddled with the thought that it would be a good idea to temporarily shift some of the rangers from the asbestos forest, i.e. Inyo and Mono, to help out.

''As I was away from headquarters most of the summer, the Shanghaiing summons seldom caught up with me before the fire was out.

''The first time I was roped in, believe it or not, was in the middle of January. It is hard for anyone not familiar with the mountains of Southern California to picture a forest fire at this season of the year, but, during years of no rain, the relative humidity hits bottom and the forest fuel is dry as the inside of a bag of unopened potato chips.

''This fire started from sparks from a chimney. It was fanned by the parched Santa Ana Winds and went roaring through the San Bernadino Mountains.

''Simpson, Guy Way and I were the handcuffed volunteers, drafted from the Inyo. We went in Simpson's car. We reported in at the Forest Service headquarters. Before we had time to indulge in the fraternal pleasantries, we were hustled out and down the street to a vacant lot where several efficient looking rangers and state fire wardens were rounding up and trying to hold a lot of unhappy potential fire fighters; sort of the 'press gangs' of His Majesty's Navy in the bygone years. A big, red-haired, red-faced forest officer detailed us to guard the lot and, noting that we didn't have guns, (previously, when fighting fire, we left our guns behind deeming that an extra canteen of water was much more necessary.) 'Well, anyhow, knock the hell out of any of the *@*!! that try to get away,' and sailed down the street to join the other 'recruits.'

''Apparently, the word had spread, and the streets were devoid of manpower because few were added after we started our patrolling.

''Trucks were backed up to the lot and a ranger addressed the assembled group in a stentorian voice of Spanish. I think the gist of the harangue was that it was their patriotic duty to fight the fire and those that tried to evade this duty would be sent to the snow encrusted parcels of Siberia or, perhaps, buried without benefit of priest or clergy.

"Their names were jotted down and they were loaded on the trucks and whisked away.

"I had used many and devious methods to collect fire fighters, but those southern rangers had the most efficient and productive methods although, at the present time, their system would produce protest marches that would block every street in Los Angeles.

"We were whisked out to the fire camp on the edge of a portion of the burned out fire line.

"The camp was in the process of getting a little order out of the beautiful turmoil and confusion of truck loads of newly arrived, uncooperative Chicanos, mountains of tools, bedding, food, etc., scattered.

"Simpson, because of his abilities, was designated to establish some order out of the surrounding chaos. Way, too heavy to work on the fire line, was to take over the distributing of tools, bedding, etc. I was sent to report to the Fire Boss who, after a little questioning, gave me the job of scout. I was given a clip board with paper and maps of the immediate area with instructions to climb to the most advantageous points and draft in the progress of the fire with notes on the topography, forest cover and passable access route to the various points of the fire line. I was given an energetic, young forest guard to relay my work to the Fire Boss. My efforts must have been O.K. as, on each fire after that, scouting was my job.

"The days were cool and pleasant, except for the wind, but the nights would freeze the reproductive glands off a simian cast in brass. There was not enough bedding and about the only relief was hot coffee. One night, Way insisted that I use his blankets while he shivered over a camp stove.

"Fire lines were jumped and the fire roared on for about five days when the winds changed, bringing damp air from the ocean, and the fire went out on its own accord.

"The last fire I was on was one to be remembered!! It was on the Santa Ana River Watershed. The topography was so steep, the squirrels wore climbing spurs just to get up and down the slopes and lizards carried hatchets to get through the brush. It was so hot that the wind felt like a blowtorch. It was on this fire that I found out what real fear was. We lost several men: A Forest Service Foreman and his crew were mopping up the fire line up a steep slope when the fire jumped the line behind them and came leaping up the unburned side of the line. The Foreman called to the crew to follow him into the burned area which was still hot, but three of the crew panicked and dashed into the unburned area and tried to race the fire. The fire won!! It made a sobering effect on the whole camp to see those bodies go by lashed to pack mules.

"Roy Boothe and I were together from the Inyo on this fire. He explained to me what courage was, physical courage. I suppose every man has it in one way or another and it assumes forms from risking of life to dying in the effort. It is usually rewarded by admiration, praise or the pinning on of medals.

"Moral courage is usually unnoticed or condemned! A forest fire is a good example:

"Let us say a large fire is burning out of control. The logical place to backfire from is a wide highway, a river or some natural firebreak, but to backfire from there means burning of much improved property or valuable watersheds. It takes lots and lots of moral guts to order the setting of a backfire that will burn out homes, and installations of friends or influential people.

"This fire posed the above problem, the Santa Ana River was the only safe line to backfire from. Roy Boothe, and several others, begged to backfire from there, but the ones in authority said, 'No,' they could stop it along the ridge. Whether they were afraid to burn out the improvements and watershed or maybe they really thought that they could stop it at the ridge 'quien sabe.' The next day, the fire jumped the fire line on the ridge, burned out the watershed and improvements, crossed the Santa Ana and roared out of control for several days until the weather changed and it went out of its own accord.

"I was never fire boss on a 'big' fire, but have often wondered if I would have the courage to set a backfire if I thought it was necessary, where it would burn out homes and would cause my being condemned and criticized.

"One very nice silvery tinkle in my memory of this fire was the hot, luxurious bath that followed. I was never a fastidious glob and could skip bathing with a conscience as pure as an unborn babe, but after a week's accumulation of smoke, ashes, dust and sweat, when I oozed into a big bath tub of hot soapy water, aahhhhh.

"Not that we wanted a car, but because everyone else had one, we weakened about the first of 1927. It was a 1923 Chevrolet roadster. The poor, little stinker! I shudder to think what, in my ignorance, I asked of it: I had driven dad's Model T Ford a little with the ulcer producing horror of anyone riding with me and never could drive a car with any idea of how things would transpire. To this day, after over sixty years of questionable maneuvering of an automobile, unless I have six car lengths, in trying to parallel park, I usually end up with my hind end on the sidewalk and my front feet out in the street.

"Our first trip is pretty well impressed on my mind: Up Westgard Pass, then out on a wood road to mark some pinon for a permittee cutting firewood, then down to the Springs at the south end of Deep Springs Valley. We arrived after dark and the wind was having tantrums. We ate cold supper in the car, then the problem of making our bed. It was impossible to hit the ground with a blanket. The sand and dust filled wind kept the blankets flapping straight out. We solved it by Helen sitting on a corner of the bed while I anchored it with rocks, then working around the edges piling stones as we went, until we had it under control and crawled in.

"The sun was coming up with a soft warm benevolent smile on his face as I sat up in bed and gouged the sand out of my eyes, nose, mouth and ears. The wind had gone with its tantrums elsewhere. I took a bucket, went down to the spring, washed and threw a few buckets full over my gritty carcass and took my first good look around. I couldn't believe it. Less than 100 yards above, on an out thrust of rock in perfect silhouette, stood a beautiful ram apparently much impressed in my ablutions. My only view of mountain sheep before was short glimpses of some that were rapidly getting out of sight. Helen had never seen one. I slithered back to camp and got Helen out of bed. I grabbed our camera, then, with Helen, clad in her nightie and shoes, we moved back to the spring. The ram had not moved and we soon spotted two ewes and their lambs. We watched them for an hour, snapping all of the film in our camera. Because of the distance and their blending with the rocks, the pictures were mostly duds. We did get one good shot of the ram against the sky line.

"The object of the side trip to the springs was to look for tracks to see if the mountain sheep were still watering there.

"Back to our trip. We drove up to the north end of Deep Spring Valley, then over the ridge and down into Fish Lake Valley, then up Fish Lake Valley making side trips up into the White Mountains to mark wood to be used to cook quicksilver out of cinnabar from a mine. Then up Chadovich Creek as far as there was a semblance of a road to check on a special use for a pipe line, then back to Deep Springs Valley. Just off of the road where Willow Creek crossed, there was a tempting, green sward. After driving all day over terrible roads in a contraption that didn't have horse sense, I threw the car into low and headed through brush to that little meadow, killed the engine by now throwing it out of gear and made camp. Next morning, we leisurely loaded the Chevy and found that the battery was deader than the mummy of Ramses II.

"Late that day, we flagged down a miner from Tonopah who said that he would deliver the battery to the Forest Service in Bishop the next morning. Then we waited!! It was the first time I had the opportunity to really look my new acquisition over. I suppose on that trip, I had done everything wrong that anyone could think of and many things no one would have thought of, but, the only thing that my uneducated eye could find wrong was that the wooden spokes in one of the front wheels were alarmingly loose. I solved this problem by whittling out wooden wedges and driving them between the spokes where they entered the hubs.

''Three days later, Simpson arrived with my charged battery and a few caustic remarks. Those were beautiful days. Nothing that had to be done. We botanized, hunted arrowheads, bathed twice a day and read. Helen had brought Ruskin's 'Crown of Wild Olives' and 'Queen of the Air.' What wonderful thoughts Ruskin had. Obviously, he had never been exposed to the gritty side of living which, I believe would have changed his philosophy. To put it another way, his opinion of wild roses would have been different had he been bucked into the bramble of them instead of just observing and smelling them.

''To stretch our grub, after many misses, I managed to shoot a jackrabbit, but he was so 'high' on the flavor of sage and rabbit brush he had been living on, that we couldn't swallow a bite. Pickles and Hogan thought it was ambrosia and cleaned up every smidgen of it. They had a glorious time on a rodent control project, enthusiastically, but unsuccessfully trying to dig out the numerous antelope chipmunks and kangaroo rats.

''Naeda was on the way. The termination of the first episode of our life together was starting to rumble. Helen could still make trips with me by car or horse and cart, but pack trips into the back country with me were ended. It was not the same.

''We had bought a ranch in Loomis which took most of our savings, plus an $8,000 note. We were trying to get it set out in fruit trees. Naeda was born in San Francisco August 20, 1928, and it was still harder to be away from home.

''About this time, we heard that the Reclamation Service was opening reclaimed land on Tule Lake for homesteading, so Alcorn, Jack, and I drove up to look it over and filed on three separate pieces chosen by Mr. Alcorn. We then forgot it as we were told that there were many applicants for each of the 28 parcels and only qualified farmers would be considered. To our surprise, in about six months, we had a notice that we had been granted our first choice so, in September 1929, I signed my resignation from the Forest Service.

''Never has it been so hard to write my name on a piece of paper and I must admit, it was splotched with tears.

"If St. Peter, in a careless moment, allows me to pass through those Pearly Gates, my hope is that it will be to again be District Ranger on the White Mountain—Big Pine District with Roy Boothe as my supervisor.

"With old Buck prancing under me and Helen beside me, Fossil Face, Winnaduma and Cactus trailing behind us, mountains around us, the fresh, cold wind in our faces—that will be our Nirvana, Kismet" (Brainerd).

Chapter Eight
The Outdoors

These are true stories relating to hunting, fishing, and trapping. In 1928, I recorded the food habits of the Sage Grouse. Our family had rabbit drives to save the apple trees and alfalfa fields. This information should be interesting to the outdoors man, wildlife biologist, and those interested in nature.

All of us Alcorn children grew up with an awareness of our surroundings. Most of the birds we saw were identified. An eagle soaring overhead was seen and admired. It sure is a beautiful sight to watch an eagle soar. I remember watching an eagle near Red Mountain Lake, he was high in the air. All of a sudden, he folded his wings and came straight down, like a bullet. He was almost down when he spread his wings and turned skyward. He generated enough speed to shoot himself straight up in the air. He made these dives down and up several times.

We watched the ground to tell us the story of the past. If a track of an animal or man was seen we made a mental note of the size of the track, which direction it was going, and how fast. Father could tell a buck deer from a doe, by looking at the size and shape of the track and how far it sank into the ground.

We used to do a lot of deer hunting, so naturally we walked a lot. Father said to Lloyd, "Lets go for a walk." They started up a hill and Lloyd was lagging behind. After waiting for him for quite a while, father said, "Well, you are not going to make it, so you better spend the night up here. Back yourself up to a tree, so nothing will get you." Needless to say, Lloyd made it down the hill, because there was no way he was going to stay the night up there alone.

Like normal kids we often ran or jogged everywhere. Physically we were in good shape, making it easier for us boys to go up into the Sierras and hunt.

When hunting we often looked for fresh buck tracks and following the tracks until we saw the deer.

Dad once said to me, "I want to show you something." It was in the middle of November. Dad showed me where some coyotes had killed a good sized deer in the south alfalfa field. Earlier that day dad had been horseback riding below Red Mountain and saw some deer tracks that headed easterly toward the alfalfa field. The field was about half a mile away from where he first saw the tracks. He decided to follow them. Riding along he saw some coyote tracks following the deer. The deer and the coyotes' tracks were digging into the dirt from running. Dad said he could see where one of the coyotes was running to one side and the other coyotes were running behind. Father showed me the tracks where they entered the alfalfa field and one deer got separated from the rest. The coyotes had taken the deer down with a struggle. We could see where the deer had gotten up and ran about fifteen or twenty more jumps, but by then, the coyotes had him down again. The deer was partially eaten. The next morning we returned to the field to see what had happened to the carcass, but there was nothing left. Coyote tracks were everywhere, they had completely eaten or carried away all of the deer.

Father went fishing at Birch Lake the day before deer seasoned opened. He was amazed to find a buck deer that had been shot, gut cleaned, and stashed under some bushes. He took a piece of rag from his pocket and stuffed it up the nostril of the deer. Arriving at the lake, father saw a man fishing. He wondered if this man had shot the deer. The following afternoon, on opening day of deer season, father saw the same deer hanging in a tree at the man's house that had been fishing at the lake. There were several neighbors standing nearby admiring the big buck. Father smiling walked over and said, "Wow! That is a nice buck, when did you shoot it?" The man proudly announced that he had killed it that very morning on opening day. Father said, "That's funny, I saw a deer that looked exactly like this one when I was at Birch Lake yesterday. It was hidden under some bushes, and I stuffed a rag up its nostril, so I could positively identify it." Father reached up the nostril and pulled the rag out. Father later said, "You should have seen the expressions on their faces."

I shot my first deer when I was eleven years old, it was one that father had wounded. I was twelve years old when I killed my second deer. Glen Moser and I went hunting on Birch Mountain. We saw a forked-horn buck run over a nearby ridge. I jumped off my horse and ran over to the ridge and fired two shots with my thirty-thirty and killed the deer. At the time, I thought, boy, I sure fired my gun before Glen had a chance to shoot. Years later it dawned on me that Glen had held back to give me a chance at that deer.

Ray Alcorn shot this deer, September 3, 1926, in Coyote Meadows. The horns are still in velvet.

When hunting with Lawson Brainerd in Coyote Meadows on September 3, 1926, I shot at a big old buck a couple of times with my twenty-five caliber Remington pump. Lawson was only a couple hundred yards away from me. He came over and asked, "Did you get him?"

I replied, "I don't know."

He said, "Let's look." We found his tracks and saw some fresh blood, so we started tracking him. Within a short distance we found the deer. We left the deer overnight at the Coyote Meadows cabin and brought it down the next day on horseback. It was a huge four point deer that weighted 198 pounds. The deer still had the velvet on it's horns. I still have the mounted deer head.

I do not recall the reason for my hunting alone one day, but I remember driving to the base of the mountains and walking to Stickers Bench. I walked above the timberline and shot a forked horn deer about three in the afternoon. I was on the north slope, rather than going back to Stickers Bench, I figured it would be best to go down Red Mountain Creek, because it would be shorter. Boy, was that a mistake!! The deer was too heavy for me to carry, so I had to drag it. Resting near an old rock slide I contemplated on which way to go. Nearby, I saw a boulder about four feet in diameter, I thought with a little push, it would start rolling down the hill. Using a tree limb as a pry, I got the boulder started. It sure was spectacular the way it went down the hill. The boulder hit something, bounced in the air, stirred up the dust and made a terrible commotion. I thought, ''Good gosh, if there is anybody else on this mountain, it would scare them to death, to hear the commotion and see something like that fly through the air.'' After the boulder went down the mountain, I realized the only possible route for me down the mountain was to cross the rock slide. It was extremely difficult. I still wonder how I made it across without starting another rock slide. I worked and worked trying to get the deer down. Finally, I got myself and the deer to the car, by then it was nearly two o'clock in the morning. Father and a group of men were out looking for me. Everyone was worried about my safety.

Father and Lloyd went deer hunting at Birch Mountain. They left the horses at the bottom of Lloyd's Gulch. We named the gulch after Lloyd because he always killed his deer there. They saw a big buck track and decided to follow it. Father said, ''Maybe, we can catch him on the top; I don't think he can go over the ridge because the other side is too steep.'' With difficulty they reached the top and were surprised to see the deer walking along the ridge toward Yosemite. The deer being to far away, they gave up the chase. Father and Lloyd looked down and could not believe how high they were. They did not think Birch Mountain was higher then the Tinnemaha Mountain. We called the Tinnemaha Mountain the Black Mountain because of its color. Since they were on the top of Birch Mountain, they wanted to leave something. Lloyd had recently received a Bronze medal for a marathon race in Bishop which had his name stamped on the back. This medal was put under one of the rocks on the very top of the mountain. Lloyd recently (1991) remarked, ''I wonder if that medal is still there?''

Frequently people from Los Angeles and other cities of Southern California would go hunting with father. Most of them had a hard time keeping up with him due to the high altitude and lack of physical strength. They often overexerted themselves. Father told us about his friend who went with him on a hunting trip. This man was huffing and puffing, pretty soon he sat down on

Left to right, Lloyd Alcorn, Harry Alcorn, Mr. Harris and Glen Moser with the deer that Lloyd shot.

a rock and started to vomit. He was feeling terrible and mumbled, "And you call this a sport?"

* * *

Summer picnics with friends and family were common. Fishing was the favorite sport for our family and friends. These gatherings were held at various creeks where we fished. Most of the trout we caught were "pan size," which were nine to twelve inches in length. After several hours of fishing, the trout were cleaned and ready to be cooked in a skillet over an open fire. We first fried home cured bacon. The fish were then dipped in flour and put into the skillet of hot bacon grease. We removed the center of the spinal bone with rib bones attached in one piece. This gave us two separate pieces of boneless meat to eat.

Along with the fish, we usually had fried potatoes. The potatoes were peeled, sliced, and fried in bacon grease to a golden brown. These feasts were often finished by eating a big slice of mother's apple pie with several spoonfuls of homemade ice cream.

J.H. Kispert, Steve Kispert, Ray Alcorn and Harry Alcorn with trout from Owens River.

When I was a young boy several adults kiddingly asked me if I carried angle worms in my mouth while I fished. I wondered what it would be like, so one day, I put a large angle worm into my mouth. The worm crawling inside my mouth gave me an odd feeling. I decided right then and there that I was not going to carry my fishing worms in my mouth.

I went on a camping trip with a man by the name of Mr. Hurtell, who was an agricultural commissioner. He was like the modern day 4-H club leader. He took a group of us boys on horseback up Bishop Creek and over Piute Summit. We made a big loop around and came back. We were gone eleven days. He took eleven loaves of bread, along with various other food to eat. Well, us boys gobbled up one loaf of bread in about five minutes and wanted more. The food had to be rationed; us boys complained about being hungry. We probably were not as hungry as we thought we were, but we still had lots of fun.

* * *

Earl and Ray Alcorn did a lot of fishing in the Tinnemaha Creek.

One spring, I was walking home from school along an irrigation ditch. Looking up, I saw a ripple in the water. Thinking to myself, "There must be a big fish in there." I got down on my stomach and felt around in the water. I thought to myself, "That fish will be surprised when I get a hold of him!" The water was muddy, but finally, I caught him! I squeezed tightly and lifted him out. "There is no way you're getting away," I thought. My "big catch" startled me because it was not what I planned on catching; it was a snake! Quickly letting go, back in the water it went.

Once I went fishing with Jim Moser. We walked up Big Pine Creek, above Glacier Lodge, to Thunder and Lightening Lake. We were standing on a snow bank, casting into the lake. Lo an behold, as Jim cast out, his pole flew out of his hand into the water. He said, "Boy, I have to get my fishing rod." He took his clothes off, dove in, got his fishing rod and climbed out. When Jim got out, he was as blue as the sky. I have often wondered what would have happened if he had gotten a cramp.

I remember going on horseback over Taboose Pass, elevation of about 11,400 feet, with a couple of fellows that my dad wanted me to guide. We stopped for the night at Bench Lake and there were some big trout. Some of them looked like they weighed over five pounds. We were never able to catch any of them. It seemed like anytime a fishing line was thrown out, it made a ripple in the water and scared the fish. The lake was so clear and beautiful, it did not make any difference whether we caught any; we enjoyed being there. From there we went south to Sawmill Pass and down into the valley.

Dad and I used to fish at June Lake. It was a long drive, so on opening day we left about two o'clock in the morning in order to get to the lake by daylight. The limit a person was allowed to catch was ten pounds and one fish a day. On one day, I was fly fishing and caught a seven pounder and a four pounder so, those two fish were my limit. Several times dad and I caught fish that weighed eight pounds or more. The water was so clear that we could look down and see schools of trout near the bottom. We often tried to catch them, by using salmon eggs or a little piece of worm and letting the bait float to the bottom. We put the bait on a size 12 hook. The object was to have the hook small, so the fish could not see it. If the hook was big enough for them to see, the fish would not take the bait. After hooking a fish, we did not reel it in right away. We played with the fish, until we could land it without breaking our line. Dad and us boys made many trips to June Lake.

Another time at June Lake, a friend of ours came along and started talking to dad. He said, "We have been up north to Topaz Lake. Do you want to see the fish we caught?" My gracious, he had a gunny sack full. Some of those fish were real beauties that may have weighed fifteen pounds. A fifteen pound trout is a huge fish. We always commented about the gunny sack full of fish.

Father and Ed Ober, the game warden, planted trout at Birch Lake. The only fish in this lake were what we called Tahoe Black Spotted, which had red meat. We thought they were the best eating fish we could possibly get. It was a long horseback ride to Birch Lake, but we usually caught a nice bunch of trout.

On one occasion, I went on horseback to Birch Lake. I was so eager to start fishing right away, that I didn't take my sleeping bag off the horse. Since the fish were not biting, I went back to camp. My horse had reached around and

Harry Alcorn with his Golden Trout from Red Mountain Lake.

taken bites out of my brand new sleeping bag. He tore it almost to pieces. I was very disappointed, but I learned a good lesson; always unpack your horse before leaving it.

Dad arranged to get some Golden Trout and plant them in Red Mountain Lake. A year or two later Earl and I decided we would walk up to Red Mountain Lake and spend the night. We brought back a nice string of fish. Father later caught a Golden Trout that weighted four pounds from this lake.

Near the ranch, father built a reservoir between the Hill Orchard and the Red Mountain. We made a ditch from Tinnemaha Creek to fill this reservoir with water. After planting bullhead catfish in the reservoir, we were soon catching catfish that were pretty good sized. We were surprised to find a trout that weighted about four pounds with two holes poked in it. We captured the near dead fish. My dad calculated a blue heron had speared it, but the fish had gotten away.

* * *

When we pruned the fruit trees, we hauled the limbs to the edge of the orchard and stacked them in small piles. The quail and the rabbits would run

Earl with a string of Golden Trout from Red Mountain Lake.

into the limb piles for protection. If we were hunting and couldn't find anything, we would climb up on the limbs and jumped up and down. The quail or cottontail rabbits frequently ran out, giving us a chance to shoot.

The main game on the ranch were quail. A group of men went out on a big quail hunt. When they came back, they had many quail to skin. At that time there was no limit on the amount of quail you could shoot. The first limit for quail was set in 1918.

In the year of the heavy snow in 1917, the quail became so weak from not being able to find any food, that we could catch them with our hands. Dad put them in the bunkhouse, where we kept and fed them until the snow melted. We saved quite a few quail that winter.

In the Red Mountain Lake area there were many Sierra Blue Grouse. If a person had a .22 rifle and there were three or four on one limb, you could shoot one, it would fall off, then shoot another one, it would fall off and so on. At one time I shot three from one limb.

I went Sage Grouse (Sage Hen) hunting with Lawson Brainerd. We were walking along a grove of aspen trees, when some grouse flew out from the sagebrush and I shot them. Lawson was on the other side of the trees and did not see what I shot. A few minutes later he came over and I showed him the birds. He said, "I don't like the looks of your Sage Grouse, you had better put them back in your bag." When he said that, I then realized I had killed Blue Grouse instead of Sage Grouse. I put the birds back in the game bag and nothing was said after that. That was the first and the last time I made that mistake.

We often went to Coyote Meadows and hunted Sage Hen with Lawson Brainerd. When we were there, Lawson had me examine the stomach contents of the Sage Grouse and write it down. I still have those notes, which were taken on September 3, 1928. They are as follows:

Sage Grouse
J.R. Alcorn 1928

Coyote Meadows, west of Big Pine, Inyo Co., California

Sage Grouse killed on this trip were examined and the following information was taken on the crop contents.

Six adult birds were examined and nine juvenile birds.

ADULTS

Bird 1	90% sage	1/2% ants	1/2% daisy flowers
Bird 2	99% sage	1/2% ants	1/2% lady beetles
Bird 3	85% sage	15% beetle	
Bird 4	90% sage	10% ants	
Bird 5	90% sage	5% ants	1% bulbs 4% daisy flowers
Bird 6	94% sage	3% ants	3% daisy flowers

JUVENILE

Bird 1	70% sage	30% ants	
Bird 2	45% sage	40% ants	15% beetles
Bird 3	75% sage	20% ants	5% beetles
Bird 4	49% sage	49% ants	1% beetles
Bird 5	45% sage	50% ants	5% beetles
Bird 6	50% sage	45% ants	5% beetles
Bird 7	85% sage	15% ants	
Bird 8	45% ants	45% Iris pods	10% beetles
Bird 9	50% sage	45% ants 2 1/2% beetles 2 1/2% daisy flowers	

The contents of the stomach were estimated, by volume, on all 15 birds examined.

* * *

At the ranch, we did not get a chance to shoot ducks very often. Once in awhile I went to the reservoir, near Red Mountain. I remember seeing five Redhead Ducks. I started shooting and with the very first shot I knocked one down. The rest of the ducks kept circling around, to where the others were. Darned, if I didn't finally get all five of those ducks. Boy, I really thought I had done something great. I took them home and we really had a feast.

Earthquakes were not uncommon. When the ground started shaking we usually looked up to the Sierras to see if there was dust from a rock slide or from a large boulder. It was a source of interest, to see what was happening on the mountains. I recall on one occasion after an earthquake, a large boulder was seen plummeting down the mountain.

* * *

William H. Alcorn an E. H. Ober with Mountain Sheep taken from Grape Vine Mountains in Inyo County, California.

Between November 10th and 20th about 1924, my father accompanied Mr. E.H. Ober, a game warden, Mr. William H. Blake and an Indian guide by the name of Johnnie Hunter to the vicinity of the Dry Mountains, located NNE of Saline Valley and SE of Eureka Valley near the Grape Vine Mountains in Inyo County, California. The purpose of this trip was to collect mountain sheep for Dr. W.T. Hornaday to put on exhibition at the Bronx Museum in New York City.

The party of four men spent the morning walking in search of mountain sheep. They saw one ram about one-half mile away. Making plans to stalk the sheep, Johnnie Hunter and Mr. Blake walked in a semicircle in an attempt to approach the animal from different directions. Mr. Ober and my father walked straight toward the animal. On reaching the area where the ram could be seen (which was on a point of the hill), my father shot and killed the ram sheep. One lamb was also seen on this point. The lamb had the choice of running back past my father and Mr. Ober or descending from the steep hill. The animal descended the hill with considerable difficulty, loosening many large boulders as it slid down the face of the hill. Father and Mr. Ober shot at the lamb about three times, but both of them missed. When the lamb was about one-fourth of a mile away, two golden eagles were seen descending from a considerable height toward the lamb. The first eagle attacking from behind appeared to alight on the back of the running lamb. It hesitated momentarily, then flew into the air at which time the second eagle approached from a similar angle and apparently repeated the same process. Father speculated that the eagles were sinking their talons into the back muscles of the lamb which would disable it. Both eagles repeated this about three or four times. The running lamb had reached the base of a large rock, which was about 50 feet or more in diameter. The two eagles were seen to alight on the rock. This incident was observed at about 3 P.M.. The hunters were a considerable distance from their camp, so they felt they did not have sufficient time to investigate the final outcome; however, the lamb appeared to be considerably weakened either by its long run or by the attack from the eagles.

After my father related this incident to me, I read to him the account given by Ober on pages 37 and 38 of "The Mountain Sheep in California" (California Fish and Game 1931, V. 17). He agreed that the two accounts have a similarity and that they probably refer to the same incident with different views.

* * *

There were lots of rabbits in the Tinnemaha area. If there was a large increase in population of jackrabbits and cottontail rabbits, they would eat the alfalfa as well as the bark off of the fruit tree trunks and kill the trees. In order to protect the trees, dad cut rabbit brush and tied it around each tree trunk. This brush tied up against the tree soon dried and the rabbits would not eat the dry brush. This was the only way our family could protect the apple tree bark from the rabbits.

There were some porcupines at the ranch that liked to climb the pear trees. They would do some damage if the trees were young.

I remember, dad used to have jackrabbit drives. On these drives, we would have a lot of men line up thirty yards apart. We would sweep through the area and drive the rabbits ahead of us into a net fence pen. The fence had a couple of wings that went out, for a hundred yards on each side, so that it was easier to get the rabbits penned up. Each one of us had one or two willow saplings, that was about an inch and a half in diameter and two feet long. If a rabbit tried to break through and come between us, we would throw the stick at them in an attempt to kill them. After the rabbits were penned up, they were killed.

Father would sometimes throw small rocks at rabbits at a distance of about 20 yards away and kill them. Us kids, trying to become as accurate as father, threw rocks all of the time. If we killed a jackrabbit with a rock, we thought we really accomplished something.

When something catches a jackrabbit it lets out an awful crying sound, which my father could imitate. The only way a person could tell whether it was a rabbit crying or my father was if they were watching him.

Dad had to put a netting fence around the south alfalfa field, to keep the rabbits out, but they would still dig under the fence and destroy the alfalfa. Dad would take Shep down to the field to catch the rabbits. It was interesting to watch the dog. When a rabbit started to run, Shep would chase it. The rabbits frequently ran toward the sagebrush that was on the outside of the net fence, they would crash into the fence and so would the average dog. But Shep was smart enough and trained to know that when the rabbit hit the fence, it would bounce back. He would grab the rabbit when it bounced back, and give it a shake to kill it. After that he would look for another rabbit to chase. Lloyd remembers Shep killed over fifty jackrabbits in one day.

We often trapped cottontail rabbits by placing a steel trap in their trails or at the entrance of a burrow. Others were shot. When the snow was heavy, with the aid of a dog, we could catch them by hand.

The rose bushes around the north side of the ranch always had a lot of cottontail rabbits. Father and one of the hired hands took our little Fox Terrier dog to hunt cottontails. The terrier was small enough to run under the bushes and chase the rabbits out.

One winter morning after a heavy snow storm the dog chased the rabbits out into the soft snow. The rabbits sank into the snow, so they were unable to run. The men harvested about seventy-five rabbits that day.

Every winter we had cottontail rabbits hanging on the north side of the house to eat. They were cleaned and hung, but not skinned. This was to keep them from drying out. Hanging them a foot apart allowed them to freeze and keep for weeks. In the Spring, occasionally a few young jackrabbits were eaten. Father told us we were not suppose to eat the rabbits in summer months because they could have a disease called ''Tularemia.''

We used to have a dog that only chased the rabbits a short distance. If he could not catch the rabbit right away, he would trot along and wait for another one to appear, then chase him. Dad said to me, ''All that dog was doing was testing his speed. If he was gaining on the rabbit, he would continue the chase until he caught it.'' I have also seen coyotes testing their speed against the rabbits, like our dog did.

Coyotes often dug holes under the netting fences especially in the orchards. They liked to go into the orchards and eat the fruit that fell off of the trees. Coyotes like pears, especially after they have been on the ground for awhile, because they become juicy and sweet. My dad used to set traps at the openings in the fence to catch the coyotes and rabbits.

Our family had a lot of sheep on the ranch, but we never seemed to have any trouble with coyotes killing them. Extra hired hands took care of the ewes during lambing season. The ewes were put into corrals while they had their lambs. Later, it was always a big job during the castrating and/or docking of the male and female lambs. Sometimes us boys, would round up and separate the lambs from the ewes. There would be about six people helping, but only one man would do the actual castrating. This event took place right on the ranch.

Before I became old enough to trap, Lloyd ran quite a trap line where he trapped coyotes and other fur bearing animals that could be sold for the fur. During his high school years Lloyd would look at his trap line on weekends. In those days, a person could look at his traps once a week or whenever he wanted to. Not every day or two like the laws require a person to do in more modern times. So, on the weekends, Lloyd would run his trap line and take care of the furs.

When I was about twelve years old father told me, ''You go up south of the reservoir by Red Mountain, and you will find quite a few coyote tracks. We have an old burro that is weak and will not last long, take him up there and shoot him for bait, then set a couple of traps around there." I lead the burro to the reservoir and looked for a place to shoot him. I thought to myself, "Well, if I shoot him here, I can set some traps in between those three or four openings through the sagebrush." I got my rifle loaded and fired one shot at the burro. I did not kill him with the first shot, I was too young to realize I could have killed him with one shot to the brain cavity. I shot again and killed him, then I set my traps. When I went back later to check my traps, I had two coyotes. Those were the first coyotes I ever trapped. Now this is in contrast to what I learned in later years. After I started trapping more, I learned to take a dead animal, leave it, and let the coyotes eat half of it before setting any traps. By doing that, if the coyotes ate part of the animal then they would not be as suspicious when they returned to eat the second time. Most of the time when a coyote sees a dead animal, unless they are really hungry, they are very spooky. A carcass that has been eaten on for a few days seems to attract more coyotes. If I had let the coyotes eat half of the carcass, I probably would have caught more coyotes instead of only two. At any rate I was real happy to catch those two coyotes.

Father showed me how to skin a coyote and stretch their hide so it would dry properly. Dad said a man from Colorado could skin a coyote in three minutes, so I was always trying to equal his record.

In about 1927 or 1928, I received $15.00 a piece for my coyote furs. My gracious, in that time $15.00 a piece was a lot of money.

In the fall of 1928, I set traps for coyotes and other fur bearing animals near the ranch. I drove the school bus, so my plan was to set the traps where I could see any animal in the trap from the school bus. One morning I was driving the bus to school and could see the ears of an animal in my trap. It looked like a bobcat. I'm afraid I did not think much about school studies that day. When school was out, I drove the school bus as fast as possible to get home. Then I ran the half mile to the trap and was disappointed to find a great horned owl instead of a bobcat in the trap. I let the owl loose and it flew away, unharmed.

Lloyd was working at the Standard Oil Plant near the Zurich Crossing. He set traps along the Owens River to catch mink. In one month, he got five mink in one location.

It was while living on the Tinnemaha Ranch that us Alcorn children learned the basics of nature and the outdoors. The teachings of father, Forest Ranger Lawson Brainerd, and others gave all of us an appreciation of our surroundings. It was these learning experiences that have been helpful in my life-long work as a wildlife biologist. These experiences are never to be forgotten.

BIBLIOGRAPHY

Brainerd, Lawson. The Autobiography of Lawson Brainerd. Mill Valley. Unpublished.

Bryan, Carroll G.. Genealogy and History of the Descendants of William Konkel. Detroit: Boulevard Press, 1966.

Chalfant, W.A.. The Story of Inyo. Rev. ed., Bishop, Calif.: Chalfant Press, [1980], c.1933.

Inyo Register. "Big Development Enterprise." February 9, 1911.

Inyo Register. "Established Nursery." November 30, 1911.

Inyo Register. "Sunland Department." September 5, 1912.

Shumaker, Gladys. Personal letter. 1978.